FRACTALS IN GEOTECHNICAL ENGINEERING

EXPLORATORY WORKSHOP, INNSBRUCK, 2003

9

General editor:

Dimitrios Kolymbas

University of Innsbruck, Institute of Geotechnic and Tunnel Engineering

In the same series (A.A.BALKEMA):

1. D. Kolymbas, 2000, *Introduction to hypoplasticity*, 104 pages, ISBN 90 5809 306 9

2. W. Fellin, 2000, *Rütteldruckverdichtung als plastodynamisches Problem*, (*Deep vibration compaction as a plastodynamic problem*), 344 pages, ISBN 90 5809 315 8

3. D. Kolymbas & W. Fellin, 2000, *Compaction of soils, granulates and powders - International workshop on compaction of soils, granulates, powders*, Innsbruck, 28-29 February 2000, 344 pages, ISBN 90 5809 318 2

In the same series (LOGOS):

4. Christoph Bliem, 2001, *3D Finite Element Berechnungen im Tunnelbau, (3D finite element calculations in tunnelling)*, 220 pages, ISBN 3-89722-750-9

5. D. Kolymbas, ed. (2001), *Tunnelling Mechanics, Eurosummerschool, Innsbruck, 2001*, 403 pages, ISBN 3-89722-873-4

6. M. Fiedler (2001), *Nichtlineare Berechnung von Plattenfundamenten (Nonlinear Analysis of Mat Foundations)*, 163 pages, ISBN 3-8325-0031-6

7. W. Fellin (2003), *Geotechnik - Lernen mit Beispielen*, 230 pages, ISBN 3-8325-0147-9

8. D. Kolymbas, ed. (2003), *Rational Tunnelling, Summerschool, Innsbruck, 2003*, 428 pages, ISBN 3-8325-0350-1

FRACTALS IN GEOTECHNICAL ENGINEERING

Exploratory Workshop, Innsbruck, 2003

Edited by

Dimitrios Kolymbas

University of Innsbruck, Institute of Geotechnic and Tunnel Engineering

The first three volumes have been published by Balkema
and can be ordered from:

A.A. Balkema Publishers
P.O.Box 1675
NL-3000 BR Rotterdam
e-mail: orders@swets.nl
website: www.balkema.nl

Cover:
Upper: Praxmarkarspitze (Karwendel) – mountain
in the calcareous Alps of Tyrol, height 2638 m
Lower: sand hill, height ca. 20 cm, courtesy of Prof. G. Gudehus

Bibliographic information published by Die Deutsche Bibliothek

Die Deutsche Bibliothek lists this publication in the Deutsche National-
bibliografie; detailed bibliographic data is available in the Internet at
http://dnb.ddb.de.

ISBN 3-8325-0583-0

ISSN 1566-6182

Logos Verlag Berlin
Comeniushof, Gubener Str. 47,
10243 Berlin
Tel.: +49 030 42 85 10 90
Fax: +49 030 42 85 10 92
INTERNET: http://www.logos-verlag.de

PREFACE

Self-similarity is a striking and ubiquitous characteristic of structures in geology. On the other hand, capturing of scale effects is one of the main challenges in geotechnical engineering. There, the usual approach is to extrapolate laboratory findings toward large scale problems. Whereas this procedure is passable for small grained soils, it proves unrealistic for coarse grained soils and rock, especially jointed one. As a consequence, the prediction of e.g. water inrush into a large tunnel based on the examination of rock samples and spot-wise packer tests proves to be hardly possible.

The new branch of fractal analysis encompasses both notions, self-similarity and scale effects. Therefore, it pricks up the geotechnical engineer's ears and evokes the hope that it can help to treat the related problems. A more thorough view, however, shows that the exploitation of fractal analysis in the above sense is not so straightforward. First of all, our limited range of observance over several scales inhibits to check and quantify the scale effect. At that, we lack the mathematical tools to treat problems that exhibit a continuous scale dependence, as this is the case with fractal materials.

The generous support of the European Science Foundation made possible to organise an exploratory workshop on that theme. The participants came from different disciplines, such as Geotechnical Engineering, Geology and Mathematics. It proved once again how difficult the inter-disciplinary communication is. I wish, therefore, to thank all participants for their cooperation. The workshop, being an exploratory one, cannot be expected to provide conclusive answers. It can be hoped, however, that the papers collected in this volume will help interested researchers to make further progress in this very promising realm.

I also wish to thank Mrs. Christine Neuwirt for the financial administration of the workshop and Ms. Myriam Berthold for the preparation of this volume.

Innsbruck, April 2004

Dimitrios Kolymbas

ACKNOWLEDGEMENT

The financial support of the Workshop by the following organisation is gratefully acknowledged:

European Science Foundation
1 quai Lezay-Marnésia
67080 Strasbourg cedex
France

TABLE OF CONTENTS

Scaling methods, fractals and the size effect of multiple fracture

15

Feodor M. Borodich

Fractality of crushed brittle materials: geometry or fracture mechanics?

37

Andrew Palmer

Fractal properties in rock fragmentation: result of a self similar process or consequence of a pure stochastic phenomenon? 47

Thierry Villemin, Luc Empereur-Mot

Application of the fractal fragmentation model to the fill of natural shear zones 57

David Masin

The Development of Fractal Geometries in Deformed Rocks   67

Bruce E. Hobbs and Alison Ord

3D Imaging of Jointed Rock Masses   79

Alison Ord, Fabio Boschetti, Bruce Hobbs

Fractal geometry analyses of rock fabric anisotropies and inhomogeneities 115

J. H. Kruhl, Francisc Andries, Mark Peternell, Sabine Volland

Fractals and scale effects in fractured multi-layered red beds 137

Christian A. Hecht

Fractals in Geotechnical Engineering

Dimitrios Kolymbas

Institute of Geotechnical and Tunnel Engineering University of Innsbruck, Technikerstr. 13, 6020 Innsbruck, Austria
e-mail: dimitrios.kolymbas@uibk.ac.at

1 Introduction

Design in Geotechnical Engineering is based on predictions of the behaviour of soil and rock. The traditional way of our predictions is from the small to the large. Based on the investigation of soil or rock samples in the laboratory we predict the behaviour of, say, a large dam. In many cases this step is unjustified because of a pronounced size effect. The material behaviour in the large is different than in the small. This poses a series of difficulties to be stated below. They can be mastered if we know the governing law. To the extend that geomaterials have a fractal nature, their mechanical properties depend on the size according to power laws. The validity of this assumption has to be investigated.

2 Fractals in Geomechanics

The standard approach in Geotechnical Engineering is to consider soil and rock as a continuum. Moreover, it is tacitly assumed that soils and rocks are *simple materials*. The consequence is that the deformation of a single point may reveal the entire mechanical behaviour of the material. This is the rationale of the basic procedure in Geotechnical Engineering: to extract samples from the field, to test them in the laboratory, to derive therefrom σ-ε-relations and then to proceed to computations. The standard continuous objects (lines, surfaces, bodies) have no internal structure, i.e. they are not subdivided into smaller entities. A part is similar to the whole, there is a one-to-one correspondence between the points of the part and the points of the whole and these points have identical mechanical properties. In contrast, geomaterials have a pronounced internal structure. Soils are particulate materials consisting of grains, and rock is usually jointed, i.e. crossed by discontinuities. This fact influences the mechanical behaviour of geomaterials, the simple-material-approach is not always valid.

An important issue in materials with internal geometric structure is the 'characteristic length', such as the grain diameter or the spacing of joints. If this characteristic length is small compared with the other important lengths of the considered problem, then the internal structure of the material can be somehow smeared, i.e. a surrogate simple material is introduced, the underlying technique being called homogenisation. In this sense, the granular nature of soils can be overlooked — for a large series of applications. The limits of this approach are still not clear and there are many cases where the homogenisation is not applicable (e.g. shear localisation).

3   Scale invariant grain size distributions

A peculiar kind of materials with internal structure are materials that *do not* possess any internal length. This can also be formulated as follows: Such materials possess an infinite spectrum of internal lengths. E.g., there are granular materials with no internal length. These are characterized by scale invariant grain size distributions.

Consider a cubical soil sample with mass M_0 and volume l^3. The grain size distribution $M(\delta)$ gives the mass of grains with diameter less than or equal to δ. The distribution curve $y(\delta)$ gives the ratio $y := M(\delta)/M_0$, where δ is logarithmically plotted. We denote with $\delta_y := \delta(y)$ the grain diameter which is not exceeded by $100 \times y$ percent of the sample mass. A grain size distribution is scale invariant if it does not have any internal length. This is the case if y depends only on the ratio δ/δ_{max}, i.e. $y = \eta(\delta/\delta_{max})$.

Consider a soil sample of the mass M_0 with self-similar grain size distribution. Let the grain diameters be within the range $0 < \delta < \delta_{max}$. If we truncate this distribution by removing all grains with $\delta > \delta_1$, we obtain the grain size distribution $\eta(\delta/\delta_1)$ of the remaining sample from the initial distribution $\eta(\delta/\delta_{max})$, see also Fig. 1:

$$\eta(\delta/\delta_1) = \frac{1}{\eta(\delta_1/\delta_{max})}\eta(\delta/\delta_{max}) \qquad 0 < \delta < \delta_1 \qquad (1)$$

The factor $1/\eta(\delta_1/\delta_{max})$ is to re-normalize the truncated distribution. The functional equation (1) is fulfilled by any power law where β is a real number.

$$\eta(\delta/\delta_{max}) = (\delta/\delta_{max})^{\beta} \ ,$$

We see thus that power laws ($y = \text{const} \cdot \delta^{\beta}$) can represent scale invariant (or self-similar) grain size distributions. Power laws are typical for fractals, which are sets consisting of parts similar to the whole, and which can be described by a fractional dimension. In a log-log-plot this relation appears as a straight line

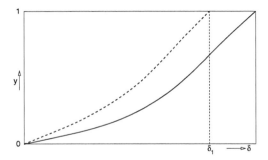

Figure 1: Grain size distribution of a sub-sample $\delta \leq \delta_1$ (dashed)

$(\ln y = \text{const}_1 + \text{const}_2 \ln \delta)$ whereas in the usual log-plot it appears as an exponential function $\left(y = \text{const}_3 \cdot 10^{\text{const}_4 \cdot \lg \delta}\right)$. In Fig. 2 are plotted in a log-log-diagramm the grain size distributions of a morainic soil and a sediment soil. It should be added, that morainic material is characterized by its geological genesis: It origins from rock broken by moving glaciers which pushed it ahead without any grain size separation. A separation can be accomplished e.g. by flowing water which exerts upon the grains drag forces that depend on their size. As a result, fluviatile sand deposits do not exhibit a power law distribution, in contrast to non-separated clastic (=fracture) sediments.

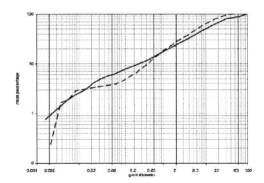

Figure 2: Grain size distribution of morainic and sediment (dashed) soils.

TURCOTTE and HUANG have listed [1] a series of fragmented objects with fractal character, i.e. power law distributions with fractal dimensions varying between 1.44 and 3.54. For morainic debris has been reported $D = 2.88$ [2]. Despite some naive models there is no convincing explanation for the fact that fragmentation leads to power law distributions.

Power laws are ubiquitous and govern many distributions in many different fields (e.g. economy). They are also called PARETO distributions according to the power law which describes the distribution of wealth. LÉVY distribution is yet another name for power laws.

Some examples of power laws are:

1. The relation between the cumulative number of earthquakes and their magnitude, known as GUTENBERG-RICHTER law

2. Relation between river length and area of the drainage basin

3. Relation between number of oil fields and oil volume

4. Distribution of faults. According to SORNETTE [6], however, the pure fractal description is too naive and more sophisticated quantifiers must introduced that reconcile the existence of hierarchical structures in many scales (see fig. 9).

4 Fractal patterns

The notion 'dimension' stems from geometrical fractals, which are geometrical objects with dimensions exceeding their topological dimension. A geometrical fractal is an object of irregular geometry whose scaling properties are described by the fractal dimension D. D can range between topologic and Euclidian dimension. E.g., a profile of a rough surface is topologically a line ($D = 1$), but is defined in Euclidian two-space, and the fractal dimension falls between 1 and 2 [5]. A fractal object may be covered by small objects (boxes). The fractal dimension D describes how the total size of the set (i.e. the number of boxes) depends on the box size. To cover a fractal curve or a fractal surface with boxes of the edge δ we need N boxes (see Fig. 3).

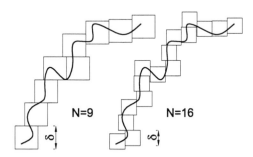

Figure 3: The number N of boxes needed to cover a curve depends on their edge length δ.

Obviously, N depends on δ: $N = N(\delta)$. The smaller δ, the larger N is. For usual (non-fractal) curves we have $N \propto \frac{1}{\delta}$ and for non-fractal surfaces we have $N \propto \frac{1}{\delta^2}$. In general $N \propto \frac{1}{\delta^D}$, where D (the so-called fractal dimension) is a fraction for fractals. E.g. the shoreline of Great Britain is reported to have the fractal dimension $D = 1.3$.

The length of a fractal curve is obtained as $L \approx N\delta = \text{const} \cdot \delta^{1-D}$. Plotting L (or N) over δ in a log-log plot gives a straight line the slope of which gives D.

The fractal nature of geological materials is revealed if we consider natural or artificial rupture surfaces. Such surfaces are self-similar in the sense that a part of them is similar to the whole: If such a surface is subdivided in smaller parts, each part looks like the original surface. This is why photographs of rock surfaces and of granular soils do not reveal the size of the objects shown, unless another object of known size (e.g. a hammer or a coin is added). Fig. 4 gives the impression of a high mountain although it shows a sand heap of ca 20 cm height.

Figure 4: Mountain or heap of sand? (Photograph: Prof. G. Gudehus)

The fact that rock surfaces are very often fractal is also revealed by the following fact: Rocks belong to the few objects that can be better painted by computers than by professional painters (cf. fig. 5 - fig. 7).

The origin of fractality in geomaterials could be seen in the fact that most of them result from fracture processes. This is the case not only for clastic sediments (clasis = fracture) but also for tectonically fissured and jointed rock (figure 9). It is interesting to note that the idea of a systematic pattern formation in joined rock goes back to GOETHE, who assumed that joints in granite result from a generic T-shaped joint pattern (fig 8).

Assertion *Fractal patterns are created by phase transitions of materials with no internal length.*

5 Fractal grains and friction angle

There is also another aspect of similarity with respect to granular soils. It is accepted that the shape of the grains plains a role in the overall mechanical behaviour of a soil (see e.g. [8]) and it is known that there are several shapes of grains. Taking into account the irregularity of the grains, how can we define similarity of two grains? Of

Figure 5: Rock surfaces in arts (painter: F. Kontoglou)

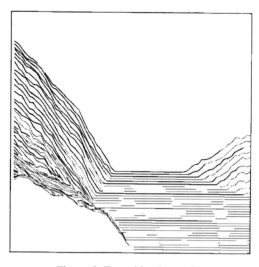

Figure 6: Fractal landscape [7]

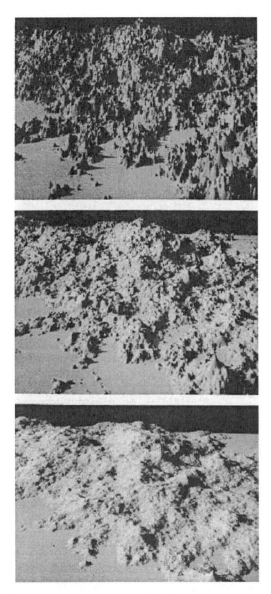

Figure 7: Fractal landscapes generated using VOSS's successive random algorithm [7]

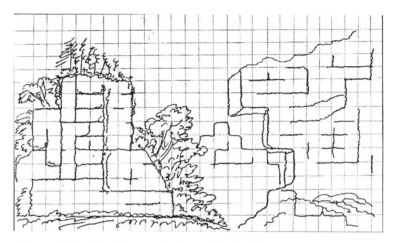

Figure 8: Draft by GOETHE [10] to explain the formation of join patterns

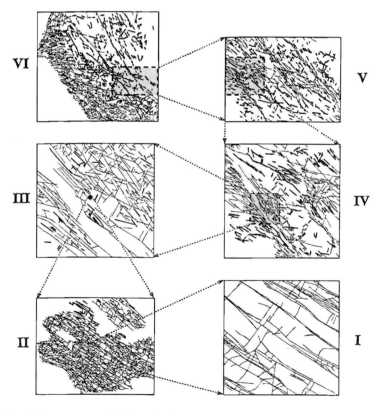

Figure 9: Series of fracture networks from the field (satellite images of the western Arabian plate) [6]

course, such a similarity can only exist in a statistical sense. Several quantities have been introduced to indicate the shape and roughnes of a grain (or its cross section). One possibility is to evaluate the FOURIER spectrum of the cross section given in polar coordinates $r(\phi)$. Another possibility is to evaluate the fractal dimension of this curve [3]. It has been found that the so obtained fractal dimension correlates with the friction angle [4].

6 Theory of V. N. Rodionov

RODIONOV assumes that the behaviour of rock is governed by inhomogeneities of, say, spherical form, the nature of which is not further specified. He only assumes that their sizes (e.g. diameters) l_i are distributed in such a way that the material does not possess any internal length. In other words, the size l_i of the largest inhomogeneity within a sample is proportional to the size of the sample. He further assumes that in the course of loading there are formed stress fields around the inhomogeneities. The deviatoric parts of the related stresses increase with deviatoric deformation (according to a constitutive equation such as $\overset{\circ}{\mathbf{T}} = \mathbf{h}(\mathbf{T}, \mathbf{D})$. RODIONOV assumes the linear elastic relation $\dot{\mathbf{T}}^* = 2G\dot{\mathbf{E}}^*$, where $\mathbf{T}^*$ is the deviatoric part of $\mathbf{T}$ and $\dot{\mathbf{E}} = \mathbf{D}$, $\dot{\mathbf{E}}^* = \mathbf{D}^*$) and decrease due to relaxation:

$$\dot{\mathbf{T}}_i^* = 2G\mathbf{D}^* - \frac{v}{l_i}\mathbf{T}_i^*.$$

Setting $\mathbf{D} = $ const and replacing tensors with scalars (e.g. components or appropriate invariants) he obtains therefrom

$$T_i^\star = 2GD^*\frac{l_i}{v}(1 - \exp(-\frac{v}{l_i}t)).$$

Based on data from wave attenuation RODIONOV assumes that v can be approximately considered as a universal constant for rock:

$$v \approx 2 \cdot 10^{-6} \quad \text{cm/s}.$$

To a particular time $t^\star$ that ellapsed from the begin of loading we can assign the length $l^\star = vt^\star$. We can then distinguish between large inhomogeneities ($l_i \gg l^\star$) and small ones ($l_i \ll l^\star$):

Large inhomogeneities: $\exp(-l^\star/l_i) \approx 1 - l^\star/l_i \rightsquigarrow T^* = 2GD^*t^\star$, i.e. deviatoric stress increases with time $t^\star$, a process which eventually leads to fracture (fragile or brittle behaviour).

Small inhomogeneities: $\exp(-l^*/l_i) \approx 0 \rightsquigarrow T_i^* = 2GD^*l_i/v$, i.e. the material around small inhomogeneities behaves as a viscous material (cf. ductile behaviour).

We see thus that the material behaviour of the considered body, in particular the distinction between brittle and ductile behaviour, depends on the value of the parameter vt/l_i . As l_i correlates with a characteristic length (size) L of the body, we can consider the parameter vt/L as determining the material behaviour. If we consider a model test simulating a prototype (e.g. the convergency of a tunnel), where model and prototype consist of the same material, then the following similarity condition must be preserved according to RODIONOV:

$$\frac{t_{model}}{t_{prototype}} = \frac{L_{model}}{L_{prototype}} \quad .$$

E.g., if we consider the convergency of two cylindrical cavities with radius R_1 and R_2, respectively, observed in the same squeezing rock, then the same convergency $u_1/R_1 = u_2/R_2$ occurs after the time lapses t_1 and t_2, respectively, where $t_1/t_2 = R_1/R_2$.

With increasing time, l^* increases until it obtains the value of L. Then the body necessarily behaves in a brittle way and failure sets eventually on.

7   Consequences of fractality for numerical simulation

Assume that we can prove that some geomaterials have, in fact, fractal properties. Of course, the range of fractality should be delimited with lower and upper bounds. The consequence will be that properties such as stiffness, strength and permeability will scale with the size of the considered body according to some power law. This will be the key to capture the size effect.

An important difficulty arises as soon as we want to take fractality into account in numerical simulations. The size effect renders the usual stress-strain relations inadequate. There are some modern constitutive relations designed for materials that possess an internal length. Such materials are called 'non-local'. They are characterized by the fact that stress depends not only of strain but also on the average of strain taken over representative finite volumes.

The averaged strain $\bar{\varepsilon}$ is defined as follows:

$$\bar{\varepsilon}(\mathbf{x}) := \frac{1}{V} \int_{\Omega} \psi(\mathbf{x} + \mathbf{s}) \ \varepsilon(\mathbf{x} + \mathbf{s}) dV \quad ,$$

where Ω is a symmetric neighborhood of the point $\mathbf{x}$ and has the volume V.

The weight function ψ is normalized:

$$\frac{1}{V} \int_\Omega \psi(\mathbf{x} + \mathbf{s}) dV = 1$$

Moreover ψ is an even function:

$$\psi(\mathbf{x} + \mathbf{s}) = \psi(\mathbf{x} - \mathbf{s})$$

Thus it can be achieved that for constant strain fields $\bar{\varepsilon}(x) = \varepsilon(x)$ holds true.

Non-local theories, where the stress is a function of $\bar{\varepsilon}$ are close related with gradient theories:

A TAYLOR-series development of $\varepsilon(\mathbf{x} + \mathbf{s})$ yields

$$\bar{\varepsilon}(\mathbf{x}) = \varepsilon(\mathbf{x}) + \frac{\partial \varepsilon}{\partial \mathbf{x}} \frac{1}{V} \int_\Omega \psi(\mathbf{x} + \mathbf{s}) \mathbf{s} \, dV + \frac{1}{2!} \frac{\partial^2 \varepsilon}{\partial \mathbf{x}^2} \int_\Omega \psi(\mathbf{x} + \mathbf{s}) \mathbf{s}^2 dV + \ldots$$

or

$$\bar{\varepsilon}(\mathbf{x}) = \varepsilon(\mathbf{x}) + c^2 \frac{\partial^2 \varepsilon}{\partial \mathbf{x}^2} \quad .$$

It should be added that constitutive theories, where the curvature κ_{ij} is taken into account are also gradient theories, because κ_{ij} is the gradient of ω_{ij}^{ink}.

A special case are the so-called gradient theories. Cosserat-theories belong to the gradient-theories. However, all these theories refer to *one* material length, whereas fractal materials do not have a material length. Moreover, the size effect of fractal materials renders any sort of finite elements inapplicable, since a fractal body may not be decomposed into small finite elements: The fractality imposes a pronounced dependence of the results on the mesh size, i.e. on the size of the finite elements. Materials with a single internal length can be numerically treated using (i) a non-local or gradient type constitutive law (and the pertinent deformations) and (ii) mesh sizes adapted to the aforementioned internal length. The latter requirement imposes extremely fine meshes, but the problem can be — at least in principle — mastered. For fractal materials, however, the difficulty imposed by fractality is a dramatic one: The assembly of the finite elements does not correspond to the body considered as a whole! It seems to be very difficult to find a way out from this dead end. Perhaps a typical length can be assigned to a given problem in such a way that the stress strain relation pertinent to this size can be used. An alternative approach could be to distribute randomly the fractal patterns with a size exceeding the one of the used mesh.

8   Open problems

How to show that geomaterials are, in fact, fractals? Or how to show that size effect is, in fact, governed by power laws? The experimental proof is hindered by the extremely narrow range of sample sizes accessible to laboratory testing (figures 10, 11, [9])

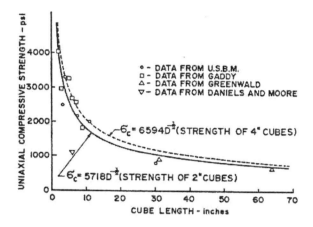

Figure 10: Size effect on rock [9]

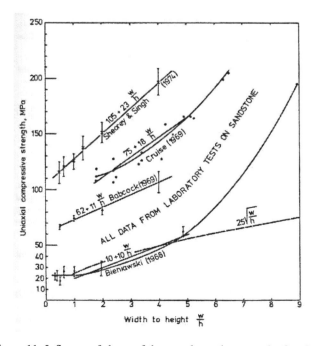

Figure 11: Influence of shape of the sample on the strength of rock [9]

Some theoretical approaches try to explain that fracture processes lead necessarily to fractal distributions. However, these derivations are rather artificial.

The only feasible way could be to assume fractality and check the consequences of this assumption, i.e. to check whether this assumption helps to quantitatively describe observations in the large.

9 Perspectives for the future

If the pattern of fissures and joints in rock is fractal, then one could predict the behaviour of a large rock mass based on the analysis of a small sample. Moreover, one could predict the water inrush into a tunnel, which is mainly governed by large open joints, based on the analysis of fissures of a small rock piece.

Bibliography

[1] D.L. Turcotte and J. Huang, Fractal distributions in Geology, In: Fractals in the Earth Sciences, Chr. Barton and P.R. La Pointe.

[2] D.L. Turcotte (1986), Fractals and Fragmentation, *J. of Geophys. Res.* **91**, 1921 - 1926.

[3] L.E. Vallejo, Fractal analysis of granular materials, *Géotechnique* **45**, No. 1, 159-163 (1995).

[4] U. Gori and M. Mari, The correlation between the fractal dimension and internal friction angle of different granular materials, *Soils and Foundations* **41**, No. 6, 17-23 (2001).

[5] S.R. Brown, Dimension of self-affi ne fractals: Example of rough surfaces, In: Fractals in the Earth Sciences, Chr. Barton and P.R. La Pointe (editors), Plenum Press 1995.

[6] D. Sornette, Critical Phenomena in Natural Sciences – Chaos, Fractals, Selforganization and Disorder: Concepts and Tools, Springer Berlin, 2000.

[7] J. Feder, Fractals, Plenum Press, New York, 1988.

[8] I. Herle, G. Gudehus (1999). Determination of parameters of a hypoplastic constitutive model from properties of grain assemblies. *Mechanics of Cohesive and Frictional Materials* 4(5): 461 - 786.

[9] Z. T. Bieniawski, Rock Mechanics Design in Mining and Tunneling. Balkema, 1984.

[10] Goethe, J. W. v.: Die Metamorphose des Granits: Substanz- und Gestaltbildung des Erdorganismus. Zusgest. u. bearb. von D. Bosse. Verlag Freies Geistesleben, Stuttgart, 1994.

Scaling methods, fractals and the size effect of multiple fracture

Feodor M. Borodich

Cardiff University, Institute of Theoretical, Applied and Computational Mechanics, School of Engineering, Cardiff CF24 0YF, United Kingdom
e-mail: BorodichFM@Cardiff.ac.uk

Abstract

Similarity and scaling approaches to mechanics of multiple fracture are under consideration. Various existing scaling techniques are discussed including classic coordinate dilation, statistical self-similarity, parametric-homogeneity, and fractal models. Size effect of fracture in quasi-brittle, polyphase samples is explained by the existence of scale-dependent growth mechanisms of the process zone. The growth of the zone is modelled as a fractal pattern of fractures. It is derived a formula for the critical tensile stress that depends on both the sample size and the size of the process zone.

1 Introduction

An analysis of published experimental studies shows that on loading a specimen of the quasi-brittle, polyphase materials like rock, concrete or ceramics having a blunt notch, the following stages of fracture may be seen: (a) micro-defects, which are dispersed randomly within the volume of a material, grow and become microcracks; (b) on further loading the microcracks accumulate within the volume of a highly stressed intact material, coalesce and a few isolated mesocracks are formed in the area near the notch tip; (c) then the intensity of mesocracking increases and the process of the coalescence of the mesocracks in turn leads to appearance of the process zone that forms of cracks, defects, and faults; (d) the macrocrack grows slowly by breaking bridges between the macrocrack tip and mesocracks of the process zone; (e) by a certain critical value of external load, an abruptly propagation of the main fracture is observed.

Experimental research shows that there is the so-called size effect in fracture of the polyphase materials. This means that a sample made of such a material exhibits different behaviour when it is of laboratory-size and when the sample size increases [3]. From the point of view of the classic linear elastic fracture mechanics (LEFM), such a behaviour is rather unusual. However, it can be explained by the presence of the damage or process zone that grows near the main crack tip during fracture of the polyphase materials [33, 10]. The process zone leads to fracture energy screening effect, i.e., the energy dissipated within the process zone is diverted from the tip of the main crack and the LEFM approach is not applicable to multiple fracture. Hence, the process of multiple fracture in these materials should be considered on several scales: micro, meso and macro. To study the process within the intermediate scale, new postulates which are outside the hypotheses of classical continuum mechanics should be employed.

It is proposed to study the size effect in materials with an extended process zone using scaling arguments of the zone growth. It is assumed that the growth mechanism of the process zone is scale-dependent [19]. In an unbounded sample or a real size structure, the zone develops until its maximum width and then it propagates along with the main crack keeping the same width. The main part of the developing process zone is assumed to be wedge-shaped, while its head (the active process zone) is bounded by an arc of a circle with the center at the main crack tip. In a bounded sample that is less than some critical size, the process zone cannot be fully developed. This is the main cause of the size effect of fracture in quasi-brittle samples.

In 1992 a model of a fractal single crack was presented by the author [13]. Both mathematical and physical fractal approaches were employed. Then it was shown [15] that the same scaling arguments are valid for fractal fracture patterns. The author has also tried to model discrete propagation of fracture using the so called parametric-homogeneous scaling [17, 19]. Here various scaling techniques and their application to fracture processes are discussed in detail.

The paper is organised as follows: in §2 various definitions of fractals are recalled. Then a classification of scalings is presented. In §3 physical fractal scaling is employed to describe multiple fracture. The active part of the zone is postulated to be a fractal cluster with multiple branches. To estimate non-linear properties of fracture the resistance curve (R-curve) approach is used. Finally, a fractal model of size effect in a sample subjected to stretching stresses is analysed.

2   Scaling and fractals

An essential element in the study of applied mathematics and mechanics is to explain physical phenomena by mathematical models. If the phenomenon is quite smooth then frequently such models lead to differential equations. The situation is more complicated when physical phenomena or objects are non-smooth. In this case the direct applications of differential models are difficult or sometimes impossible at all. However, if the phenomenon is self-similar then its behaviour can be extrapolated and predicted even if it is nonsmooth. Fortunately, there is a quite old observation that goes back to Richardson [42] and Kolmogorov [34] (see also [9, 7]). A statement about the observation can be formulated as follows: processes in a surprisingly large number of cases are, broadly speaking, self-similar on their intermediate stage when the behaviour of the processes has ceased to depend on the details of the boundary or initial conditions. Now this idea is undergoing the upsurge of interest due to the introducing the concept of fractals. Although fractals and scaling models are often considered as synonyms, there are various other scaling methods that can be used to describe multiple fracture.

2.1 Definitions of fractals

What is a fractal? The term was introduced by B.B. Mandelbrot who published a book concerning fractals in 1975 [35]. It seems to the author that Mandelbrot was impressed by a popular book about sets written for Russian high-school children by Vilenkin [47] where numerous examples of sets which are more irregular than sets considered in common textbooks on Euclidean geometry were collected. Indeed, about a quarter of all figures presented by Mandelbrot have analogs in Vilenkin's book, these include the von Koch curve and snowflake, the Menger sponge, the Cantor staircase (devil's staircase), trajectory of Brownian motion and the Peano curve. By the way, the Sierpinski carpet was shown and discussed by Vilenkin, while this classic fractal was not presented in Mandelbrot's book. Surprisingly, one could not find any definition of fractals in this book by Mandelbrot [35]. In fact, this is not a unique book about fractals where one cannot find the definition of the main term. In 1977 Mandelbrot ([36], p. 15) gave the following definition of fractal sets in a metric space (**M1**): *A fractal will be defined as a set for which the Hausdorff-Besicovitch dimension strictly exceeds the topological dimension.* However, later he [37] withdrew the original definition and suggested to use the term "without a pedantic definition". In papers concerning fractals, he gave examples in order to explain what is a fractal (see, e.g., [39]). An example of his alternative definition was cited by Feder [29], namely (**M2**): *A fractal is a shape made of parts similar to the whole in some way.* Then he said (**M3**): *Broadly speaking, mathematical and natural fractals are shapes whose roughness and fragmentation neither tend to vanish, nor fluctuate up and down, but remain essentially unchanged as one zooms in continually and examination is refined* [39]. Evidently, both M2 and M3 are non-mathematical definitions and even very smooth objects can satisfy them, e.g. smooth parametric-homogeneous functions [17]. Greenwood [30] noted that Mandelbrot is somewhat reluctant to define 'fractals' or 'fractal dimension' preferring to offer examples, while Edgar [27] said that a term without a pedantic definition cannot be discussed mathematically.

Definitely, the concept needs some clarification. Otherwise fractality is not a theory, but just a confused conglomeration of various ideas of similarity theory, iterative models and theory of measure. Giving definition of fractals as *sets with non-integer fractal dimension*, the author emphasized often the necessity to split the term in two: mathematical and physical fractals (see, e.g., [15, 18]). Confusion of these two kinds of fractals led often to various erroneous or at least unjustified conclusions (see discussion in [18]). Thus, in the author's personal opinion, fractality is not unified. It can be subdivided on the following partially overlapping areas:

mathematical fractals or fractal geometry that studies various fractal and multifractal measures and fractal dimensions, in particular Hausdorff dimension and box-counting dimension (this can be obtained by consideration the limit behaviour of covers of a set by boxes of size at most δ when $\delta \to 0$);

physical (natural, empirical) fractals which are real or numerically simulated objects exhibiting a kind of self-similarity (this is the so-called fractal behaviour) in a bounded region between upper (Δ_*) and lower (δ_*) cutoffs;

iterated sets and mappings and iterated function systems, for example the ones used for image compression; and

pretty colored pictures built by the use of iterated function systems and modern computer facilities.

These areas do not have much in common and are united through either presence of power law functions and/or iterated manner of constructions. The misconception of some vital for the practical use definitions and terms of fractal analysis, and bounds for practical application of fractal concepts were discussed earlier, see for example, discussions by a group of scientists from Jerusalem [4, 5, 11] and other discussions [40, 32, 18]. However, one can still encounter misuses of the concept. The use of a fractal as a generated term without additional explanations contributes to difficulties in understanding and reproducing previously published results or leads to their misinterpretation. Hence, a question could arise [32]: *Is there meaning in fractal analyses?*

The Jerusalem group performed an extended analysis of data published by physical reviews journals and showed that the overwhelming majority of reported physical fractals span about 1.5 orders of magnitude [4]. In other words, the average ratio of the upper (Δ_*) cutoff of the fractal law to the lower (δ_*) cutoff is only $\Delta_*/\delta_* = 10^{1.5} \approx 31.6$. Although Mandelbrot [40] explained the limited-range examples of power law correlations as unfortunate side effects of enthusiasm, imperfectly controlled by refereeing, the group argued that the limited-range empirical fractals are dominant fractals observed in nature [5, 11].

One can see that the main distinction between physical and mathematical fractals is that the power law of natural objects (empirical fractals) is observed on a bounded region of scales only, while mathematical fractals consider limits when the scale of consideration goes to zero. Hence, the scaling approach to these two kinds of fractals is not the same.

2.2 Fractal fracture

Let us imagine that a tip of a continuous crack is shielded by a growing pattern of mesocracks, which is described as a fractal cluster. We suppose that if the size $\mathcal{R}$ of the fracture pattern is less than some critical Δ_*, which is the upper cutoff for the fractal law, then its growth is self–similar in fractal sense with respect to the group of homogeneous coordinate dilations.

We suppose that the fractal dimension of the mesocrack pattern is equal to some C and is constant during the cloud growth. Analysis of published experimental data concerning fractal dimensions of profiles of fracture surfaces (D) and patterns of fracture (C) for quasi-brittle materials shows that the values of the fractal dimensions belong mainly to the following intervals: $1.04 < D < 1.33$ and $1.47 < C < 1.79$ [15]. If we use the fractional part D^* of the fractal dimension, $0 < D^* < 1$, then we can write $D = 1 + D^*$ in the case of a fractal curve and $D = 2 + D^*$ in the case of a fractal surface. Hence, the above estimations may be written as the following intervals: $0.04 < D^* < 0.33$ and $0.47 < C^* < 0.79$.

2.3 Classification of scalings

We will consider scaling methods based on dilations of coordinates. If the dilation is the same along all axes then it is a homogeneous transformation $\mathbf{x} \to \lambda\mathbf{x}$. If the dilation is different along at least two axes then it is a quasi-homogeneous (or anisotropic) transformation [2].

2.3.1 Classic scaling

Geometric similarity is the classic scaling. This means that the continuous group of homogeneous or quasi-homogeneous coordinate dilations Γ_λ is acting on the object under consideration. The concepts of homogeneous and quasi-homogeneous functions are also based on these transformations. Let us recall that the function $Q_d : \mathbb{R}^k \to \mathbb{R}$ is called a quasi-homogeneous function of degree d with weights $\alpha = (\alpha_1, \dots, \alpha_k)$ if it satisfies the following identity

$$Q_d(\Gamma_{\lambda^\alpha}\mathbf{x}) = Q_d(\lambda^{\alpha_1}x_1, ..., \lambda^{\alpha_n}x_k) = \lambda^d Q_d(\mathbf{x}).$$

Homogeneous functions H_d are a particular case of quasi-homogeneous functions when $\alpha_1 = ... = \alpha_k$. If $H_d : \mathbb{R} \to \mathbb{R}$, then $H_d(x) = cx^\alpha$ where c is a constant and α is an exponent. Hence, if $k = 1$ then a homogeneous function is a power law function.

The dimensional analysis and the inspectional analysis are based on the homogeneous and quasi-homogeneous transformations of coordinates.

Let us consider the application of dimensional analysis to fracture in geometrically similar solids. It is known that before a large real structure is build, experiments on models of reduced scale are carried out [7]. Clearly, one must know how to scale these experimental results up to the full-scale structure. Let us denote by indices M and F values related to a model and the full-scale structure respectively. If two bodies have geometrically similar shapes then the function ($f^{(M)}(\mathbf{x})$) of the shape of the first body (the model) can be transformed by homogeneous dilations λ along all axes $x_i \to \lambda x_i$ into the function ($f^{(F)}(\mathbf{x})$) of the shape of the second body (the full-scale structure)

$$f^{(F)}(\mathbf{x}) = \lambda f^{(M)}(\lambda^{-1}\mathbf{x})$$

where λ, $1 < \lambda < \infty$ is the coefficient of magnification. In the framework of LEFM, we may assume that the critical stress σ_f is determined by the following quantities: the crack length l, the critical stress intensity factor K_c, the Poisson ratio ν of the material. Applying the dimensional analysis (see, e.g. [7]), one can obtain the following rule for scaling of the critical load for quasi-brittle fracture:

$$\sigma_f^{(F)} = \sigma_f^{(M)} \frac{K_c^{(F)}}{K_c^{(M)}} \left(\frac{l^{(F)}}{l^{(M)}} \right)^{-1/2} . \tag{1}$$

However, if the model and the structure are made of the same material, then we obtain from (1) the LEFM size law

$$\sigma_f^{(F)} = \sigma_f^{(M)} \left(\frac{l^{(F)}}{l^{(M)}} \right)^{-1/2} = \sigma_f^{(M)} \cdot \lambda^{-1/2}. \tag{2}$$

Let us consider an application of the inspectional analysis to equations that are self-similar under quasi-homogeneous (anisotropic) scaling. The classic one dimensional (1D) diffusion equation is an example of such an equation (see, e.g. [48])

$$\frac{\partial}{\partial t} u(t, x) = \kappa \frac{\partial^2}{\partial x^2} u(t, x),$$

where κ is called diffusion coefficient, or molecular diffusivity. The physical dimension of κ is $[\kappa] = L^2 T^{-1}$, where T is the dimension of time and L is the dimension of length. If one seeks a solution $u(t, x)$ to the equation that is a quasi-homogeneous function of degree d of time t and space x variables with weights α_1 and α_2, respectively, then he obtains that $\alpha_1 = 2\alpha_2$. Hence, $u(t, x) = \lambda^{-d}u(\lambda^2 t, \lambda x)$. Taking $\lambda = t^{-1/2}$, one obtains $u(t, x) = t^{d/2}u(1, x/t^{1/2}) = t^{d/2}\Psi(x/t^{1/2})$ where Ψ is a function of one variable. Additionally, if the mass of diffusing material is constant and initially it was concentrated at the coordinate origin then a dimensionless form of the self-similar solution may be presented by the Gaussian distribution

$$u(t, x) = At^{-1/2}\exp[-x^2/(4\kappa t)]$$

where A is an arbitrary constant.

2.3.2 Statistical scaling or self-similarity of patterns

It is known that data in the form of a set of points, distributed in an irregular way within a planar region, arise in many disciplines. One of popular techniques of statistical analysis of spatial point patterns is the so-called distance method or the theory of the nearest neighbour. The method considers a point as the basic sampling unit and the distances to neighboring points are recorded, i.e. the distances to the first, second, ... , the kth nearest point. This technique converts a list of point coordinates to a unique data set relevant to study of the population density. It is known that if the spatial pattern is characterized by some one-dimensional probability distribution function $f_X(x)$ for the distances to the nearest point then the distribution function can be completely represented by its mean μ_X, i.e. its expected value $E(X)$,

$$\mu_X = E(X) = \int_0^\infty x f_X(x)dx,$$

and its higher central moments

$$\mu_X^{(n)} = E\left[(X - \mu_X)^n\right] = \int_0^\infty (x - \mu_X)^n f_X(x)dx, \quad n = 1, 2, \dots$$

Using the mean, the higher central moments can be made dimensionless $\mu_X^{(n)}/\mu_X$. Hence, the statistical properties of the point pattern can be characterized by a single quantity with the dimension of length, namely the average distance $< l(t) >= \mu_X$ between points, and by an ensemble of dimensionless statistical characteristics. Further we will work only with average distances of patterns and denote them just as l omitting the average sign $<>$.

It is assumed that the transformation of the point pattern is a steady-state process and it transforms with the process time statistically in a self-similar way. The self-similarity means that the distribution of the points, that is normalized by the average distance, is the same for dimensionless time, i.e. only the mean of the probability distribution changes its value while all other dimensionless central moments remain unaltered. Hence, when one looks at the images of the pattern at an initial moment t_0 and at an arbitrary moment t, he cannot distinguish them statistically if he does not know the average distance between the points. Hence, one can write $l(t)/l(t_0) = F(t/t_0)$ where F is a function of the dimensionless time. It is easy to show that $F(x) = x^\alpha$, $\alpha = F'(1)$, i.e. it is a power law function. Thus, if $t/t_0 = \lambda$ then $l(\lambda t_0) = \lambda^c l(t_0)$.

This kind of scaling was applied to describe the growth of damage in fatigue tests by Barenblatt and Botvina [8].

2.3.3 Parametric-homogeneous scaling

Parametric-homogeneity studies parametric-homogeneous (PH) and parametric quasi-homogeneous (PQH) functions, PH- and PQH-sets, and corresponding transformations. PH-functions and PQH-functions are natural generalizations of concepts of homogeneous and quasi-homogeneous functions when the discrete (discontinuous) group of coordinate dilations (PH-transformation) $(\Gamma_{p^{\alpha k}})$ where

$$\Gamma_{p^{\alpha k}} \mathbf{x} = (p^{k\alpha_1} x_1, ..., p^{k\alpha_n} x_n), \quad p > 0, \quad k \in \mathbb{Z},$$

is considered instead of the continuous group of coordinate dilations $x \to \lambda x$. The function $B_d : \mathbb{R}^l \to \mathbb{R}$ is called a parametric-quasi-homogeneous function of degree d and parameter p with weights $\alpha = (\alpha_1, \ldots, \alpha_l)$ if there exists a positive parameter p, $p \neq 1$ such that it satisfies the following identity

$$B_d(\Gamma_{p^{\alpha k}} \mathbf{x}; p) = B_d(p^{k\alpha_1} x_1, ..., p^{k\alpha_l} x_l; p) = p^{kd} B_d(\mathbf{x}; p), \quad k \in \mathbb{Z}$$

and the parameter is unique in some neighbourhood. Here $\mathbb{Z}$ is the set of integer numbers. PH-functions are a particular case of PQH-functions when $\alpha_1 = ... = \alpha_l$. To avoid a non-unique definition, the least $p : p > 1$ is taken as the parameter.

The graphs of these functions can be both continuous and discontinuous, they can also be smooth, piecewise smooth, with singular points of growth, fractal, non-fractal nowhere differentiable [16]. Smooth sinusoidal log-periodic functions and fractal Weierstrass type functions are examples of 1D PH-functions [14, 17]

$$b_0(x; p) = A \cos(2\pi \ln x / \ln p + \Phi) \quad \text{or} \quad b_\beta(x, p) = \sum_{n=-\infty}^{\infty} p^{-\beta n} h(p^n x),$$

where A and Φ are arbitrary constants and h is an arbitrary function. In particular, we can write the so called Weierstrass-Mandelbrot function

$$C(x; p) = \sum_{n=-\infty}^{\infty} p^{(D-2)n}(1 - \cos p^n x), \qquad 1 < D < 2.$$

where D is equal to the box dimension of the graph. One can see that the PH-functions may have the same global trend, while they have different local behaviours (see, e.g. Fig. 1 where graphs of two PH-functions having the same scaling parameter p and the same degree $d = 0$ are presented). It is easy to check that the PH-

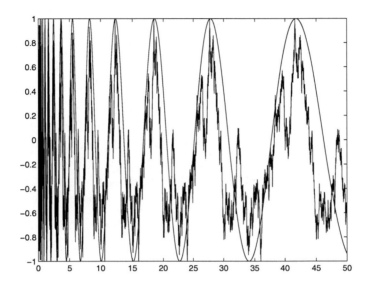

Figure 1: Smooth and fractal PH-functions with the same parameter p and the same degree $d = 0$.

functions near any point x_0 are repeated in scaling form near all points $p^k x_0$, $k \in \mathbb{Z}$. For example, if $b_d(x_0; p) = A x_0^d \sin(2\pi \ln x_0/\ln p + 2\pi k)$, then for $b_d(p^k x_0; p)$, we have

$$A(p^k x_0)^d \sin(2\pi \ln p^k x_0/\ln p) = p^{kd} A x_0^d \sin(2\pi \ln x_0/\ln p + 2\pi k) = p^{kd} b_d(x_0; p).$$

Examples of PH-sets are both smooth logarithmic spirals and such fractal self-similar sets as the Sierpiński carpet and triangle, the von Koch curve and the Cantor discontinuum. Indeed, if a part of the Sierpiński carpet with a size equal to 3 without a square $0 \le x < 1$, $0 \le y < 1$ is considered (it is the so called fundamental domain), then the whole carpet can be obtained by a PH-transformation of the fundamental domain with $p = 3$. Similar constructions can be applied to the von Koch curve and the Cantor set when the fundamental domain is $(1/3, 1]$.

Thus, the PH-sets and PH-scaling can arise in systems having a fixed scaling parameter p. If the fundamental domain is somehow filled then one can obtain the whole set by applying a PH-transformation to the fundamental domain. If the filling is fractal then the whole set is also fractal.

2.3.4   Scaling of mathematical fractals

The scaling properties of mathematical fractals are continuous and based on the following property of homogeneity for the Hausdorff s-measure $m_H(S, s)$

$$m_H(\lambda S, s) = \lambda^s m_H(S, s) \quad \text{for every } \lambda > 0, \tag{3}$$

where

$$(\mathbf{x} \in \lambda S) \Longleftrightarrow (\lambda^{-1} \mathbf{x} = \mathbf{x}_1 \in S).$$

Clearly, if a set S is a subset of λS for every $\lambda > 0$ then S is a non-fractal set. Indeed, if x_1 belongs to both S and λS for any λ then the whole ray λx_1 belongs to λS and, hence, the dimension of the set is equal to the dimension of the space X. Thus, the statement $S \subset \lambda S, \forall \lambda > 0$, where S is a fractal, may be understood only in a statistical sense. On the other hand, if there are some discrete values of λ such that $S \subset \lambda S$ is realised exactly then S is a fractal parametric homogeneous set.

The Hausdorff dimension of a set does not change under the transformation of homogeneous dilation of coordinates because the fractal measures $m_H(\lambda S, s)$ and $m_H(S, s)$ have the jump from 0 to ∞ for the same s. It is important to note that the measure $m_H(S, dim_H S)$ not always has physically meaningful values in the following range

$$0 < m_H(S, dim_H S) < \infty, \tag{4}$$

i.e. bounded, positive, and non-zero values. Instead of this, $m_H(S, dim_H S)$ may also be zero or infinite. If a Borel set satisfies the condition (4) then it is called an *s-set*. Fortunately, s-sets are very common (see for details [28]). The concept of s-sets was generalized by the author and it was applied to all physically meaningful fractal s-measures including box-counting measure. Thus, one says that a set is *D-measurable* if its s-measure has a finite positive value $m_D(S)$ for s equal to the fractal dimension D [15]. The scaling property of $m_D(S)$ is similar to the property (3)

$$m_D(\lambda S) = \lambda^D m_D(S). \tag{5}$$

The use of the mathematical fractal scaling leads often to rather complex mathematical constructions and it stipulates the introduction of new approaches and new concepts which are appropriate for mathematical description of fractal processes [7, 13, 15]. For example, if the fracture pattern or a crack is imagined as a mathematical fractal then it is more natural to attribute physical quantities to the fractal measure m_D of the considered mathematical model, rather than to the infinite length of the fractal curve, or to the infinite area of the fractal surface. In 1992 the author proposed to use the *specific energy absorbing capacity of a fractal surface* $\beta(D^*)$.

Physically, $\beta(D^*)$ gives the amount of elastic energy spent on forming a unit of the fractal measure m_D. The physical dimension of $\beta(D^*)$ is

$$[\beta(D^*)] = FL/L^{2+D^*},$$

where F is the dimension of force and L is the dimension of length. Using the $\beta(D^*)$ and the scaling properties of the fractal measure, one can obtain the scaling formula for the surface energy of a D-measurable fractal crack or a pattern of multiple fracture [13, 15].

2.3.5 Scaling of physical fractals

Since there is no canonical definition of physical fractals, let us consider an example, namely a profile that may be imaged on a computer screen as a union of points (pixels) of the size δ_*. Then we can obtain a computerised estimation of the number of pixels $N(\delta_*)$ forming the line and lying inside a circle or a square box of size $\mathcal{R}$ centered at a point $\mathbf{x}$. The line is a fractal cluster with dimension D if the number of its parts $N(\delta_*)$ satisfies the so-called number-radius relation

$$N(\delta_*) \approx (\mathcal{R}/\delta_*)^D, \qquad \delta_* < \mathcal{R} < \Delta_*. \tag{6}$$

for scales $\mathcal{R}$ such that $\delta_* < \mathcal{R} < \Delta_*$, where δ_* and Δ_* are upper and lower cutoff for fractal law.

The value of D is estimated as the slope of linear growth of $\ln(N(\delta_*))$ against $\ln(\mathcal{R})$. We can consider another variant of the technique, namely the region under consideration is divided into discrete cubes with side length δ_*. To get the value D of the dimension, the smallest number $(N(\delta_*))$ of E-dimensional cubes needed to cover the set within the E-dimensional region of size $\mathcal{R}$ is counted. If some "mass" $M(S_*)$ is assigned to the elementary particle S_* of the size δ_*, then instead of the s-measure m_s of the cluster S used in the definition of mathematical box dimension, one has the "mass" of the cluster $M(S) = N(\delta_*)M(S_*)$ within a region of size $\mathcal{R}$. If the profile has fractal properties then both the relation (6) and a corresponding scaling property for the fractal mass can be obtained

$$M(S) = \lambda^D M(S_*), \quad \lambda = (\mathcal{R}/\delta_*) \tag{7}$$

repeating the procedure of estimation for different values of $\mathcal{R}$. Note the similarity between the scaling properties (5) and (7). Using the scaling properties of physical fractals, scaling formulae were obtained for the fracture energy of a single fractal crack or a fractal pattern of multiple fracture [13, 15, 19, 20].

2.3.6   Other scalings

Studies of geophysical processes in block media showed that some of the processes possess hierarchical self-similarity [43].

It was found that both the earthquake energies and their spatial distributions have hierarchical self-similar properties [44, 45]. In particular, if the earthquake events have hierarchy of their energies then for every earthquake event of an energy level K, there are about 3.5 events of an energy level $(K - 1)$. This idea is very close to the idea of the existence of a fixed scaling parameter [16, 17]. These phenomena with a fixed scaling parameter observed by Sadovskii and his co-workers [43, 44, 45] were discussed later by Suteanu *et al.* [46].

It is necessary to note that the above natural PH-phenomena are not deterministic and, therefore, the scaling parameters p of the phenomena have stochastic nature. There are various probabilistic scaling approaches to fracture, for example, a scaling with probabilistic transition from one level to another one [1] or renormalization approach to a probabilistic hierarchical structure when its elements are statistically uniform and have random strength. Such structures are used for studying a variety of natural non-linear phenomena, including strength of fibrous composites, fracture of solids, seismic activation, as well as fire propagation. The failure process starts from the lowest structural level of a system and then propagates from level to level due to subsystem failures and the consequent load redistribution and transfer. The system collapse occurs when the failure process envelops the entire system. However, the probabilistic scaling approaches are out the scope of this paper. For further details on this kind of scaling see a recent paper by Onishchenko [41]).

Even scaling methods that are based on coordinate dilations may be more complicated than the above mentioned methods. For example, it was shown above that a set can have a PH-scaling globally, while locally it is a fractal. However, we cannot apply local scaling to a deterministic fractal.

Let us consider another example, namely the Cantor middle-third set (the Cantor discontinuum D_C) on $(1, 2]$. This is a PH-set with $p_1 = 3$, i.e. if $x \in D_C$ then $p_1^k x \in D_C$, $k = -1, -2, \ldots$. On the other hand, it is evident that one can consider the semi-interval $(1, 2]$ as a fundamental domain for a set on the whole $\mathbb{R}$ with $p = 2$. Hence, a set may have a PH-scaling globally with a scaling parameter p, while locally it has a PH-scaling with a scaling parameter $p_1 \neq p$.

Let a spatial pattern be presented by a set of points with average distance l. Let us consider these points as centres of fractal subsets whose sizes is less than l. These fractal subsets can represent faults or cracks. Evidently, if the pattern is developed in time in a self-similar manner then one observes a power law for distances $l(t)$, while spatial pattern is a fractal.

Finally, let us mention an interesting approach to spatial patterns of faults having local fractal properties. To describe both the spatial and length distributions of fracture networks, one may use a double power law that was introduced by Davy and his co-workers (see, e.g. [26, 23])

$$n(l, L) = cL^{D_M} l^{-\alpha}$$

where $n(l, L)dl$ is the number of fractures having a length between l and $l + dl$ in a box of size L, D_M is the mass fractal dimension of fracture barycenteres, a is the exponent of the frequency distributions of fracture lengths, and α is a fracture density term. Recently, much theoretical and experimental attention has been directed to studying the above mentioned and various other scaling phenomena and approaches to their description (see, e.g. [12, 23]).

2.3.7   Self-affi ne scaling

The concept of self-affine fractals is often mentioned in studies of roughness of fracture surfaces. A number of experimental investigations claim that profiles of vertical sections of real surfaces are statistically similar to themselves under repeatedly magnifications; however, the profile should be scaled differently in the direction of nominal surface plane and in the vertical direction. Self-affine mapping on a plane means a one-to-one mapping such that images of any three collinear points are collinear in turn. However, in the literature devoted to fractals, self-affine mapping means usually its particular case, namely the quasi-homogeneous (anisotropic) coordinate dilation, $x' = \lambda x$ and $y' = \lambda^H y$ where H is some scaling exponent.

The concept of self-affine fractal was introduced by Mandelbrot [38]. However, again he did not give any strict definition of the concept and preferred to give some examples of self-affine fractals, such as the Brownian motion $V_{0.5}(t)$ or a fractional scalar Brownian motion $V_H(t)$ [37, 29]. Of course, one could define *a self–affine fractal set as a set that is invariant from the statistical point of view under quasi-homogeneous (anisotropic) scaling*. However, the Mandelbrot's statements about dimensions of the self-affine fractals would not follow from the definition. These statements could be formulated as follows. Each version of fractal dimension for self-affine fractals has a local and a global value, separated by a crossover. Hence, there are both local and global scaling for self-affine fractals.

It is necessary to note that the above statements about self-affine fractals were not given in a rigorous manner that is traditional for mathematicians. This leads to doubts concerning the validity of the concept. The author is not convinced by the above arguments in favor of the statements about global and local fractal dimensions. Further, we have seen above that a self-similar solution to the 1D diffusion equation is a quasi-homogeneous function with weights $\alpha_1 = 2\alpha_2$. It is known that the

Bachelier-Wiener process (the Brownian motion) can be linked mathematically with the diffusion equation. In fact, diffusion reflects the average behavior of particles at the macroscopic level that are described by the Brownian motion at the microscopic level. Hence, it is not a surprise that $V_{0.5}(t)$ has self-affine properties (anisotropic scaling) and $V_{0.5}(t)$ is statistically identical to $\lambda^{-1/2}V_{0.5}(\lambda t)$. However, the author does not see any connection between this kind of scalings and fractal dimensions of the so called self-affine fractals.

Finally, if f is a one-to-one map of X onto Y and f, f^{-1} both are continuous then f is called a homeomorphism. If in addition f, f^{-1} satisfy Lipschitz the conditions on $S \subset X$, and $f(S) \subset Y$, respectively then f is called a Lipschitz homeomorphism. Any differentiable mapping is Lipschitz [2]. It is known that both the Hausdorff dimension and the box dimensions keep the same value under the action of Lipschitz homeomorphism. Hence, both fractal dimensions are the same under action of quasi-homogeneous transformation. It seems to the author that claims about self-affine scaling are not justified.

3   Fractal model of multiple fracture

The multiple fracture consists of a cascade of interacting defects of various length scales, hence direct application of LEFM to such objects, disordered by defects, is impossible [6]. It was assumed that the polyphase materials under consideration are quasi-brittle. The concept of quasi-brittle materials supposed that there is a narrow layer of non-elastic deformations near the fracture surfaces. During the crack propagation, both this layer and the new fracture surface absorb energy.

3.1   Fracture energy of the process zone

To study multiple fracture, the following postulates will be employed: (i) the width h of the layer of inelastic deformation near any fracture surface is constant and it is the same for mesocracks and the macrocrack; (ii) material of every cube of size h centered in a point of fracture surface absorbs the same quantity of energy g_f; (iii) no new microcracks arise in the compressed region; (iv) the process zone develops during loading, the size of process zone (its width and length) depends on both the fine structure of material and the stress field; (v) the jump of the main crack tip leads to a relaxation of the stress field in some domain behind the tip, and the micro- and mesocracks situated in the domain will not grow further, therefore, the process zone can be separated on an active S_A and a passive part S_P, the active part only is essential for propagation of the main crack; (vi) the stress concentration regions near the mesocracks of the periphery of the active domain are sources for growing of new microcracks.

Let us cover all mesocracks of the active part of the process zone by cubes of the size h centered in points of fracture surfaces. Then the average amount $< W >$ of absorbed energy can be calculated as

$$< W >= g_f N(h) + \text{const} \tag{8}$$

where $N(h)$ is the minimal number of the cubes in the cover.

The idea of covering the cracks of the process zone by cubes in order to calculate the amount of absorbed energy is very close to the methods of applied fractal geometry. Evidently, this approach can be used if the width of the plastic layer is small with respect to the average distance between mesocracks.

It was shown that fractal dimension of a fracture surface can correlate with fracture energy only in quasi-brittle materials with very narrow plastic zones. If fracture surface exhibits fractal features, however the width of the plastic zone h is about the upper cutoff of the fractal law, i.e. $h \sim \Delta_*$, then fracture energy is mainly related to the work done within the non-fractal zone of inelastic deformation of the crack and there is no correlation [15, 13]. Analysis of experimental data showed that for metals $\Delta_* \simeq 0.1$ mm [22], while $h \simeq 0.4$ mm [21], i.e., $\Delta_* \leq h$. Hence, the fractal properties of fracture surface are not essential for fracture energy of ductile materials. This conclusion is in agreement with experimental data [31]. On the other hand, the experimental studies show [49, 25] that for polyphase materials the upper cutoff for fractal law for fracture patterns is $\Delta_* > 1$ cm, i.e., $\Delta_* \gg h$. Therefore, the fractal properties of the pattern of mesocracks could characterize the fracture energy of the materials when the fractal properties of the main crack surface are not essential [15]. Thus, the experimental evaluation of the dimension C of the fracture pattern is more important for fracture mechanics than the evaluation of the dimension D of the fracture surface.

It is assumed in our model that (i) the growth of the process zone results in a continuous growth of the main crack; (ii) the fractal scaling is applicable to describe the beginning of the self-similar growth of the process zone; (iii) while a fracture pattern is already fully developed, i.e., when the width of the process zone reaches some critical size w_*, the pattern picture is repeated. This means that the active part of the process zone is invariant with respect to continuous shifting.

Let us use two systems of coordinates, namely (X, y) and (x, y) where $x \equiv X - l$. If the width of the sample is L and the initial notch length is l then the sample points are situated in the range $0 \leq X \leq L$. The process zone is assumed to consist of two parts, namely a wedge-shaped part with some angle 2α at the wedge vertex, i.e. at $x = 0$, and the second part of the zone is described as a segment of a circle. Hence, if the main crack has the tip at a point $x = x_0$ then for $0 \leq x \leq x_0$, it is wedge-shaped, while it is described by a segment of a circle of radius $w_c(x_0) = x_0 \sin \alpha$ for $x \geq x_0$.

It is assumed that the active part of the process zone is situated within the segment and the pattern of mesocracks within the segment is a physical fractal of dimension C for scales $\delta_* = h < \mathcal{R} < w_c(x_0)$. Substituting (6) into (8), one can calculate the average amount of energy $< W >$ absorbed by mesocracks of the process zone

$$< W > \sim g_f(w_c/h)^C + \text{const}, \qquad C = 1 + C^* \tag{9}$$

where $w_c(x)$ is the current width of the zone.

The fracture energy $G_F = dW/dx$. Hence, differentiating (9), one obtains

$$G_F \sim g_f(1 + C^*)(w_c/h)^{C^*} \cdot w_c'(x)/h = g_f(1 + C^*)(w_c/h)^{C^*} \sin \alpha/h, \tag{10}$$

for the growing pattern of mesocracks and

$$G_F \sim g_f(1 + C^*)(w_*/h)^{C^*} \sin \alpha/h = \text{const} \tag{11}$$

for the fully developed pattern of mesocracks. First, the formulae similar to (10) and (11) were obtained for a single fractal crack [13] and then for a fractal pattern of multiple fracture [15]. Bažant [10] introduced a very convenient notation G_f for a macrofracture energy, i.e. $G_f = g_f(1 + C^*)(w_*/h)^{C^*} \sin \alpha/h$ and rewrote our formulae for a single fractal crack in the following form

$$G_F = \begin{cases} G_f(x/\Delta_*)^{C^*}, & x \leq \Delta_*, \\ G_f, & \Delta_* < x. \end{cases}$$

These formulas for unbounded samples are in qualitative agreement with experimental results concerning the behaviour of fracture energy of large concrete samples [24].

In the case of nonlinear fracture of an elastic body with a cut that is under a uniaxial tensile stress σ perpendicular to the cut, the fracture energy criterion can be written as

$$G(\sigma, l) = R(l)$$

where $R(l)$ is the R-curve (the resistance curve) of the sample that is equal to its fracture energy G_F.

It is known that the elastic energy release rate $G(\sigma, l)$ is a linear function of the crack length l. In the LEFM it is assumed that $R(l) = G_c = \text{const}$. However, in nonlinear fracture mechanics, the energy $R(l) = G_F$ is used where G_F is an increasing function of the crack length. The criterion says that if $G < R$ then the crack does not propagate and if $G > R$ then the crack is unstable and its propagation is catastrophical.

In an elastic solid under a uniaxial tensile stress σ, the energy released after creating a straight-through planar crack perpendicular to the stress direction can be estimated if we follow the idea that there exist a domain where the stress field relaxes then we can state the principle of independence of released energy on crack trajectory: *the total released energy in an elastic solid does not depend on crack trajectories if the crack shape is fixed in some small regions at the crack tips and the trajectory variations are within the domain of the stress field relaxation.* [15]. It follows from this principle that the total released energy does not change if there are some micro- and mesocracks within the domain of the stress field relaxation. Thus, $G(\sigma, l)$ is still a linear function of the size l.

3.2 Size effect

At the beginning of the process, the R curve is equal to G_F. However, when mesocracks ahead of the tip of the main crack reach the sample boundary, the value of the fracture energy decreases from G_f to the fracture energy of the ultimate link g_f and (11) is not valid any more. R on this stage may be approximately described as a decreasing linear function $-Ax + B$ with some constants A and B [19, 20]. Thus, for a large sample with fully developed process zone or for the full-size structure of the size $L^{(F)}$ with a notch of the size $l^{(F)}$, the resistance curve has the following form

$$
R^{(F)} = \begin{cases} G_f(x/\Delta_*)^{C^*}, & h < x < \Delta_*, \\ G_f, & \Delta_* < x < x_{max} - w_*, \\ (G_f - g_f)(x_{max} - x)/w_* + g_f, & x_{max} - w_* < x < x_{max}. \end{cases}
$$

where $x_{max} = L^{(F)} - l^{(F)}$ and $\Delta_* = w_*/\sin \alpha$.

We assume that the main cause of the size effect is that the process zone cannot be fully developed in a bounded model of a real size construction. So, if the size of the model is small and $L^{(M)} - l^{(M)} < \Delta_* + w_* = \Delta_*(1 + \sin \alpha)$ then the growth of the process zone stops at some size w_c when cracks ahead of the tip of the main crack are reaching the sample boundary $L^{(M)} - l^{(M)} = \Delta_c + w_c = \Delta_c(1 + \sin \alpha)$. Hence, its width cannot reach the value w_*. Thus, the resistance curve does not have the intermediate stage of steady-stage development

$$
R^{(M)} = \begin{cases} G_f(x/\Delta_*)^{C^*}, & h < x < \Delta_c, \\ (G_f(\Delta_c/\Delta_*)^{C^*} - g_f)(x_{max}^{(M)} - x)/w_c + g_f, & \Delta_c < x < x_{max}^{(M)}. \end{cases}
$$

If the line G is tangential to R at some length $l = l_0 + x_c$ then $G(\sigma, l_0 + x) > R$ for $x > x_c$ and the crack starts to propagate catastrophically. However, if $x_c > \Delta_*$ in the full-scale structure then the catastrophical propagation starts at $x = \Delta_*$ because $G(\sigma, l_0 + x) > R$ for $x > \Delta_*$. Evidently, the catastrophical propagation starts at $x^{(M)} = \Delta_c$ in a model. Therefore, we can write the following formulae for the fracture stresses

$$k(\sigma_f^{(F)})^2(l^{(F)} + \Delta_*) = G_f$$

and

$$k(\sigma_f^{(M)})^2(l^{(M)} + \Delta_c) = G_f(\Delta_c/\Delta_*)^{C^*}.$$

Thus, if $L^{(M)} - l^{(M)} \leq \Delta_* + w_*$ then we obtain the size effect law of fracture of polyphase samples under tension

$$\sigma_f^{(F)} = \sigma_f^{(M)} \cdot \sqrt{\frac{l^{(M)} + \Delta_c}{l^{(F)} + \Delta_*}} \left(\frac{\Delta_*}{\Delta_c}\right)^{C^*/2} = \sigma_f^{(M)} \cdot \lambda^{-1/2} \sqrt{\frac{1 + \Delta_c/l^{(M)}}{1 + \Delta_*/l^{(F)}}} \left(\frac{\Delta_*}{\Delta_c}\right)^{C^*/2} \tag{12}$$

where $\lambda = L^{(F)}/L^{(M)} = [L^{(F)} - l^{(F)}]/[L^{(M)} - l^{(M)}]$.

Let us analyse the size effect law (12). It can be seen that if the model is not small and $L^{(M)} - l^{(M)} \geq \Delta_* + w_*$ then $\Delta_* = \Delta_c$ and the size effect law (12) goes to the fracture stress law (2) of the LEFM.

If there is no initial notch, i.e. $l^{(F)} = l^{(M)} = 0$, then the process zone starts to develop from some initial flaws and, hence, we obtain from (12)

$$\sigma_f^{(F)} = \sigma_f^{(M)} \cdot (\Delta_c/\Delta_*)^{(1-C^*)/2}. \tag{13}$$

On the other hand, we have $L^{(M)} = \Delta_c + w_c = \Delta_c(1 + \sin\alpha)$ or $\Delta_c = \lambda L^{(F)}/(1 + \sin\alpha)$. Hence, if we consider a specimen of size $L^{(F)} = \Delta_* + w_* = \Delta_*(1 + \sin\alpha)$ then (13) transforms to

$$\sigma_f^{(F)} = \sigma_f^{(M)} \cdot \lambda^{-(1-C^*)/2}.$$

In general case, $\Delta_* = \kappa L^{(F)}$ where κ is a parameter. Hence, we have

$$\sigma_f^{(F)} = \sigma_f^{(M)} \left(\frac{\kappa(1 + \sin\alpha)}{\lambda}\right)^{-(1-C^*)/2}.$$

4 Conclusions

Scaling methods apply wherever there is similarity across many scales. The similarity may be found in geometry or in the evolution of a process. In contrast to fluid mechanics, scaling in solid mechanics has been narrowed for a long time to just the equivalence of nondimensional parameters characterising the problem, while other scaling techniques were neglected. However, a wealth of empirical rules usually expressed as power laws, have been established indicating that various processes of solid mechanics are self-similar. In particular, it follows from the above analysis that scaling methods may be very effective in description of fracture processes and the fractal scaling is just a particular case of these methods.

We have seen also that scaling laws allow bridging of the scales in multiple fracture. Assuming that the process zone cannot be fully developed in a bounded model, the size effect formula (12) was obtained.

In general, the processes of multiple fracture cannot be described by the LEFM and other parameters related to the size and structure of the process zone. The verification of the postulates describing the evolution of process zone would require more complete data of mesocrack distribution within the active part of the zone.

Bibliography

[1] Allègre, C.J., LeMouel, J.L. & Provost, A.: Scaling rules in rock fracture and possible implications for earthquake prediction, *Nature* Vol. 297, pp. 47–49, 1982.

[2] Arnold, V.I.: Ordinary Differential Equations, Berlin, Springer-Verlag, 1991.

[3] Atkins, A.G.: Scaling laws for elastoplastic fracture, *Int. J. Fracture*, Vol. 95, pp. 61–65, 1999.

[4] Avnir, D., Biham, O., Lidar (Hamburger), D. & Malcai, O.: On the abundance of fractals, Fractal Frontiers, Eds.: M.M. Novak and T.G. Dewey, World Scientifi c, Singapore, pp. 199–234, 1997.

[5] Avnir, D., Biham, O., Lidar, D. & Malcai, O.: Is the geometry of nature fractal? *Science,* Vol. 279, pp. 39–40, 1998.

[6] Barenblatt, G.I.: Some general aspects of fracture mechanics, Modeling of Defects and Fracture Mechanics, Ed. G. Herrmann, Springer-Verlag, Wien, pp. 29–59, 1993.

[7] Barenblatt, G.I.: Scaling, Self-Similarity, and Intermediate Asymptotics, Cambridge University Press, Cambridge, 1996.

[8] Barenblatt, G.I. & Botvina, L.R.: Self-similar nature of fatigue failures and damage accumulation of fracture, *Mechanics of Solids* Vol. 18, pp. 161–165, 1983.

[9] Barenblatt, G.I. & Zeldovich, Ya.B.: Self-similar solutions as intermediate asymptotics, *Ann. Rev. Fluid Mech.*, Vol. 4, pp. 285–312, 1972.

[10] Bažant, Z.P. & Planas, J.: Fracture and Size Effect in Concrete and Other Quasibrittle Materials, CRC Press, Boca Raton, 1998.

[11] Biham, O., Malcai, O., Lidar, D.A. & Avnir, D. Fractality in nature: Response. *Science*, Vol. 279, pp. 785–786, (1998).

[12] Bonnet, E., Bour, O., Odling, N.E., Davy, P., Main, I., Cowie, P. & Berkowitz, B.: Scaling of fracture systems in geological media, *Reviews Geophys.*, Vol. 39, pp. 347–383, 2001.

[13] Borodich, F.M.: Fracture energy in a fractal crack propagating in concrete or rock. *Doklady Akademii Nauk (Russia)* Vol. 325, 1138–1141, 1992, English transl. in: Trans. (Doklady) Russian Akademy of Sciences: Earth Science Sections, Vol. 327, pp. 36–40, 1992.

[14] Borodich, F.M.: Similarity properties of discrete contact between a fractal punch and an elastic medium, *C. r. Ac. Sc.* (Paris), Ser. 2, Vol. 316, pp. 281–286, 1993.

[15] Borodich, F.M.: Some fractal models of fracture, *J. Mech. Phys. Solids*, Vol. 45, pp. 239–259, 1997.

[16] Borodich, F.M.: Renormalization schemes for earthquake prediction, *Geophys, J. Intern.*, Vol. 131, pp. 171–178, 1997.

[17] Borodich, F.M.: Parametric homogeneity and non–classical self–similarity, I. Mathematical background, *Acta Mechanica*, Vol. 131, pp. 27–45, 1998.

[18] Borodich, F.M.: Fractals and fractal scaling in fracture mechanics, *Int. J. Fracture*, Vol. 95, pp. 239–259, 1999.

[19] Borodich, F.M.: Self-similar models and size effect of multiple fracture, *Fractals*, Vol. 9, pp. 17–30, 2001.

[20] Borodich, F.M.: Scaling in multiple fracture and size effect, Analytical and Computational Fracture Mechanics of Non-Homogeneous Materials, Ed.: B.L. Karihaloo, Kluwer Academic Publishers, Dordrecht, pp. 63–72, 2002.

[21] Botvina, L.R., Ioffe, A.V. & Tetyueva, T.V.: Effect of the zone of plastic deformation on the fractal properties of a fracture surface. *Metal Sci. Heat Treatment*, Vol. 39, pp. 296–300, 1997.

[22] Bouchaud, E., Lapasset, G. & Planès, J.: Fractal dimension of fractured surfaces: a universal value? *Europhys. Lett.*, Vol. 13, 73–79, 1990.

[23] Bour, O., Davy, P., Darcel, C., & Odling, N.E.: A statistical scaling model for fracture network geometry, with validation on a multiscale mapping of a joint network (Hornelen Basin, Norway), *J. Geophys. Res.*, Vol. 107, # 2113, 2002.

[24] Brameshuber, W. & Hilsdorf, H.K.: Influence of ligament length and stress state on fracture energy of concrete, *Engineering Fract. Mech.*, Vol. 35, pp. 95–106, 1990.

[25] Chelidze, T., Reuschlé, T. & Guéguen, Y.: A theoretical investigation of the fracture energy of heterogeneous brittle materials, *J. Phys.: Condens. Matter.*, Vol. 6, pp. 1857–1868, 1994.

[26] Davy, P., Sornette, A. & Sornette, D.: Some consequences of a proposed fractal nature of continental faulting, *Nature*, Vol. 348, 56–58, 1990.

[27] Edgar, G.A.: Measure, Topology, and Fractal Geometry, Springer-Verlag, Berlin, 1990.

[28] Falconer, K.J.: Fractal Geometry: Mathematical Foundations and Applications, John Wiley, Chichester, 1990.

[29] Feder, J.: Fractals, Plenum Press, New York, 1988.

[30] Greenwood, J.A.: Problems with surface roughness, Fundamentals of Friction: Macroscopic and Microscopic Processes, Eds. I.L. Singer and H.M. Pollock, Kluwer, Boston, 57–76, (1992).

[31] Ikeshoji, T. & Shioya, T.: Brittle-ductile transition and scale dependence: fractal dimension of fracture surface of materials, *Fractals* Vol. 7, pp. 159–168, 1999.

[32] Jelinek, H.F., Jones, C.L. & Warfel, M.D.: Is there meaning in fractal analyses? Complex Systems '98. Eds.: R. Standish et al., pp. 144-149, 1998.

[33] Karihaloo, B.L.: Fracture Mechanics and Structural Concrete, Longman, London, 1995.

[34] Kolmogorov, A.N.: The local structure of turbulence in incompressible fluids at very high Reynolds numbers, *Doklady Ac. Sc. USSR*, Vol. 30, pp. 299-303, 1941.

[35] Mandelbrot, B.B.: Les Objects Fractals: Forme, Hasard et Dimension. Flammarion, Paris, 1975.

[36] Mandelbrot, B.B.: Fractals: form, chance, and dimension, W.H. Freeman, San Francisco, 1977.

[37] Mandelbrot, B.B.: The Fractal Geometry of Nature, W.H. Freeman, New York, 1982.

[38] Mandelbrot, B.B.: Self-affine fractals and fractal dimension, *Phys. Scripta*, Vol. 32, pp. 257–260, 1985.

[39] Mandelbrot, B.B.: Fractal geometry: what is it, and what does it do? *Proc. R. Soc. Lond. A*, Vol. 423, pp. 3–16, 1989.

[40] Mandelbrot, B.B.: Is Nature fractal? *Science*, Vol. 229, pp. 783, 1998.

[41] Onishchenko, D.A.: Scale-invariant distributions in the strength problem for stochastic systems with hierarchical structure, *Dokl. Akademii Nauk*, Vol. 368, pp. 335–337, 1999.

[42] Richardson, L.F.: Atmospheric diffusion shown on a distance-neighbour graph. *Proc. R. Soc. London A*, Vol. 110, 709-737, 1926.

[43] Sadovskii, M.A.: On natural fragmentation of rocks, *Dokl. Akademii Nauk*, Vol. 247, pp. 829–832, 1979.

[44] Sadovskii, M.A., Golubeva, T.V., Pisarenko, V.F. & Shnirman, M.G.: Characteristic rock dimensions and hierarchy properties of seismicity, *Izv. AN SSSR Fiz. Zemli*, # 2, pp. 3–15, 1984.

[45] Sadovskii, M.A. & Pisarenko, V.F.: Seismic process in block media, Moscow, Nauka, 1991.

[46] Suteanu, C., Zugravescu, D. & Munteanu, F.: Fractal approach of structuring by fragmentation, *PAGEOPH*, Vol. 157, pp. 539–557, 2000.

[47] Vilenkin, N.Ya.: Stories about sets, Nauka, Moscow, 1965, English transl. Academic Press, New York, 1968.

[48] Zeldovich, Ya.B. & Myshkis, A.D. Elements of Mathematical Physics, Nauka, Moscow, 1975.

[49] Zhao, Y., Huang J., & Wang, R.: Fractal characteristics of mesofractures in compressed rock specimens, *Int. J. Rock Mech. Min. Sci.*, Vol. 30, pp. 877–882, 1993.

Fractality of crushed brittle materials: geometry or fracture mechanics?

Andrew Palmer

University Engineering Department, Trumpington Street, Cambridge CB2 1PZ, England
e-mail: acp24@eng.cam.ac.uk

Abstract

Sea ice is technologically important because it can apply large forces to fixed structures such as offshore production platforms. The maximum ice force per unit contact area is found to be proportional to area to the power $-1/4$, and this appears to be a consequence of the very low fracture toughness of ice. Ice that crushes against a structure breaks into thousands of fragments, and the fragments may be a valuable clue to crushing processes. The size distributions of fragments of crushed brittle materials have a fractal dimension close to 2.50, and this appears also to apply to ice. The paper presents a mechanical theory of crushing that leads to the observed dimension. It presents data on the fine structure of the distribution of contact force, which appears also to be fractal.

1 Introduction

This work was stimulated by research on ice mechanics. Sea ice is encountered in several technological contexts: bridge piers in ice-covered rivers, icebreaking ships, offshore production platforms. When we think of ice in Austria we immediately think of glaciers, but glaciers are not typical. In many valley glaciers ice creeps, though in Alaska and elsewhere there are fast-moving surging glaciers whose flow appears to be dominated by fracture. At the higher strain rates we meet in technology the ice almost invariably fractures, except when the ice is moving very slowly, and except also in process zones close to the tips of cracks. Ice fractures readily because its fracture toughness is extremely low, 0.1 MN/m$^{3/2}$, about one-tenth that of glass.

Sea ice is far from homogenous [6]. Figure 1 is a highly simplified schematic. On the surface is usually snow. Under that comes the layer that froze first. That layer has small randomly-oriented grains, but further down the grains are much larger and columnar with their c-axes aligned with the current. The salt tends to be excluded at grain boundaries, and drops out of the ice in brine channels. The bottom is in equilibrium with seawater, but the top is usually much colder.

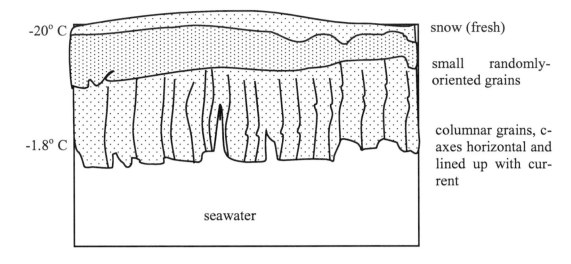

-20° C — snow (fresh)

small randomly-oriented grains

-1.8° C — columnar grains, c-axes horizontal and lined up with current

seawater

Figure 1: Schematic of sea ice structure

Suppose we make a naive estimate of the force that Arctic sea ice 2 m thick might apply to a production structure 100 m across, so that the projected area in the ice movement direction is 200 m². If we measure the strength of ice in a laboratory we find a compressive strength of about 5 MPa. So if we idealise ice as a plastic material, we find a possible ice force of 1000 MN (multiplied by some indentation factor not much greater than 1). The ice force governs design.

Creep is the governing mechanism only at very low ice speeds, and the ice force is then straightforward to calculate by finite element or finite difference methods. At higher speeds fracture governs. The ice breaks into fragments, forms a rubble field in front of the structure, and pushes round it. In most fracture mechanics contexts one is concerned with a single crack in an otherwise intact structure, but here there are thousands upon thousands of cracks, reaching from horizon to horizon.

Sanderson [4] plotted ice force per unit area against contact area. Figure 2 is a simplified version of his diagram; both scales are logarithmic. The left-hand cloud of data represent laboratory-scale tests, and the second group field tests with a contact area about 1 m². Happily Nature carries out tests on a larger scale than we could do. In Kennedy Channel between Greenland and Canada lies Hans Island. In July, large masses of sea ice several km across drift down the channel, and occasionally they hit the island. Accelerometers on the ice measure the deceleration, the mass can be estimated, and Newton's law gives the ice force. The third cloud represents the data from Hans Island. The fourth cloud has a lesser status, because it is indirect: it represents the ice forces that have to be put into numerical models of the Arctic Ocean ice sheet to make the models fit observations.

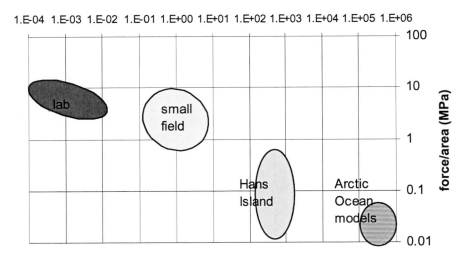

Figure 2: Sanderson pressure-area diagram [4]

Figure 2 shows that the ice forces measured at Hans Island were much smaller than would be predicted from the laboratory and small field tests. Part of the explanation comes easily from dimensional analysis. If it is assumed that the ice force P (dimension F) depends simply on the contact area A (dimension L^2) and the fracture toughness K (dimension $FL^{-3/2}$), the only dimensionally consistent form the relationship can take is:

$$\frac{P}{KA^{3/4}} \; = \; \text{constant} \tag{1}$$

It follows that if K is a material constant the ice force per unit area P/A must be proportional to $A^{-1/4}$, which is roughly what is observed, though it should be added that a few researchers do not accept the dependence of P/A on area.

2 Fractal distribution of fragment sizes

Observation makes clear that more than one mode of fracture occurs when ice breaks against a structure. One clue to what is happening might be the size distribution of broken fragments. Consider a distribution of particle sizes where $N(a)$ is the number of fragments larger than a diameter a. A fractal dimension D is defined by

$$\frac{N(a)}{N(b)} = \left(\frac{a}{b}\right)^{-D} \tag{2}$$

Turcotte [5] collected data on crushed brittle solids, and found that for many of them D is constant over a surprisingly large range, a factor of three orders of magnitude in diameter. Remarkably, the fractal dimension is almost always very close to 2.5. The exceptions are materials where sorting processes such as sediment transport in water or air have been involved.

Benoit Mandelbrot said that the greatest challenge was to understand the physical basis of fractals. Several hypotheses for a fractal dimension close to 2.5 have been put forward, and Xu [7] describes the theories and the experimental evidence relating to ice. Some of the hypotheses are based on geometry, some on Weibull statistics, and some on mechanics [2].

Imagine a hierarchy of different sizes of fragments, called here *orders*, extending from infinitely big to infinitely small. Two numbers characterise the hierarchy. s is the ratio of linear dimensions between one order and the next. m is the ratio of numbers of fragments between one order and the next. The properties of the hierarchy are summarised in the table below:

order		r	$r-1$	$r-2$	
diameter	...	d_r	d_r/s	d_r/s^2	...
total number	...	n	mn	m^2n	...
force to fracture one fragment	...	$\alpha K d_r^{3/2}$	$\alpha K d_r^{3/2}/s^{3/2}$	$\alpha K d_r^{3/2}/s^3$	...
number intersecting plane	...	n	mn/s	m^2n/s^2	...

and the fractal dimension is

$$D = \frac{\log m}{\log n} \tag{3}$$

Imagine the fragments jumbled up at random, and draw a plane intersecting the assemblage. Suppose the plane to intersect n fragments of order r and diameter d_r. There are more fragments of the next order down, but they are smaller, and so the number of intersections of the next order by the same plane is m/s. By the dimensional argument used to derive equation (1), the force B_r required to break one brittle particle of order r must be related to its diameter d_r in the following way:

$$B_r \;=\; \alpha K d_r^{3/2} \tag{4}$$

Consider a plane that intersects n fragments of order r, and calculate the force across that plane, assuming that every fragment is on the point of breaking. The force is:

$$\alpha K d_r^{3/2}\left(\ldots + n + n\left(\frac{m}{s^{5/2}}\right) + n\left(\frac{m}{s^{5/2}}\right)^2 + n\left(\frac{m}{s^{5/2}}\right)^3 + \ldots\right) \tag{5}$$

The ratio of successive terms is $m/s^{5/2}$. If that ratio is less then 1, the series converges to the right; if it is more than 1, the series diverges to the right. There is plainly something special about $m/s^{5/2}$ being 1. From equation (3), that is exactly the value that makes the fractal dimension 2.5.

Imagine now that one of the fragments of order r breaks into m fragments of order r-1. Out of those m new fragments, only m/s are intersected by the plane, and the force across the plane becomes

$$\alpha K d_r^{3/2}\left(\ldots + (n-1) + (n+1)\left(\frac{m}{s^{5/2}}\right) + n\left(\frac{m}{s^{5/2}}\right)^2 + n\left(\frac{m}{s^{5/2}}\right)^3 + \ldots\right) \tag{6}$$

and is unchanged if $m/s^{5/2} = 1$ and D is 2.5. The same result holds if any number of fragment of any order break into fragments of the next order down, and if fragmentation continues to subsequent orders.

If $m/s^{5/2}$ is less than 1, D is less than 2.5, and the additional fragmentation described above leads to a reduction in the force across the plane. If $m/s^{5/2}$ is greater than 1, D is greater than 2.5, and the additional fragmentation leads to an increase in the force across the plane.

Though there is no clear physical reason why all the terms in (5) should be equal, the above argument suggests that the fractal dimension of 2.5 corresponds to a kind of stable crushing in which the force is constant. However, an objection is that the model is founded on the hypothesis that all the fragments are on the point of breaking. That is not consistent with much recent research on the internal mechanics of granular systems. Numerical experiments using the discrete element method (DEM) shows that most of the force is transmitted by relatively few force chains, along which a force is transmitted from particle to particle, whereas the particles on either side of the chain carry relatively little force, and are therefore far from breaking. Their role seems to be to stabilise the chains. However, it should be added that DEM experiments generally consider particles that are spherical or elliptical, and that their diameters are the same or nearly the same. The angular fragments produced by fragment fracture may behave rather differently.

3 Experimental results from ice mechanics

It is implausible that the ice force can depend only on the contact area. If there is a wide contact, much wider than the ice thickness, it seems unlikely that the ice on one side of the contact can 'know' that it is part of a much wider contact that also includes the other side.

Another experiment throws light on that objection. A tactile sensor can measure the detailed distribution of force across a contact area. In one series of experiments, a platen covered with a sensor that divided the contact into 1936 segments each 1.9 mm square was driven into the edge of a plane sheet of ice. The contact force on each segment was measured 135 times a second. Figure 3 shows the force distribution in frame 176 (selected at random) of test 93. Figure 4 shows the force distribution in frame 181, 36 ms later.

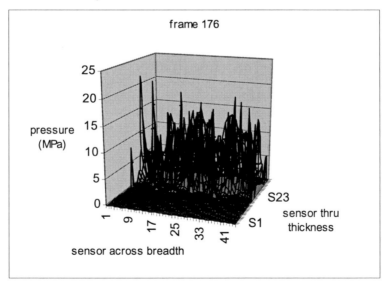

Figure 3: Contact pressure distribution across platen, frame 176 (data courtesy of Dr. D. Sodhi)

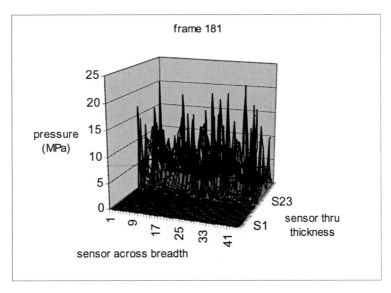

Figure 4: Contact pressure distribution across platen, frame 181 (data courtesy of Dr. D. Sodhi)

The distributions show pronounced irregular peaks. Comparison between the frames shows that the distributions are qualitatively similar, but that the peaks are in completely different places at diffrent times. This is confirmed in Figure 5, which plots the difference between frames 181 and 176: it can be seen that the differences are of the same order of magnitude as the peaks in either frame.

The peaks are called high-pressure zones (HPZs) or hot spots. They transmit most of the force. A simple idealised two-dimensional model of an HPZ considers it as line-like, concentrated along a line midway between the upper and lower surfaces of the ice. The corresponding force is then proportional to $K\sqrt{t}$ per unit distance along the line (where K is again the relevant fracture toughness and t the ice thickness (Dempsey et al. [1], Palmer and Dempsey [3]). The ice force for a contact area of breadth D is then proportional to $KD\sqrt{t}$, which is of course still consistent with a dimensional analysis . The nominal contact pressure (total force)/(Dt) is proportional to $t^{-1/2}$. That is consistent with what is observed in field-scale measurements from load panels on the Nordstromsgrund lighthouse in the northern Gulf of Bothnia, though the details are still confidential.

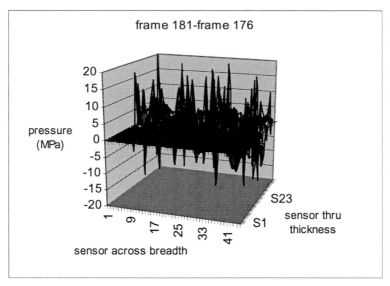

Figure 5: Contact pressure distribution across platen, difference between frame 181 and frame 176
(data courtesy of Dr. D. Sodhi)

An alternative three-dimensional model of concentrated point HPZs arrives at the same conclusion, if the HPZs are supposed to be uniformly distributed along the contact breadth and if all of them reach their force maxima simultaneously, so that the forces they generate individually are strongly correlated along the contact breadth [1, 3]. However, both the tactile sensor experiments and the field-scale measurements at Nordstromsgrund suggest that force maxima may be associated with high correlation across the contact breadth, but that correlation varies with time and is sometimes much weaker.

Recent research on ice mechanics has shifted back towards deterministic models, and away from an emphasis of the role of fractals. The pendulum has perhaps swung too far. Further research towards and understanding of the fractal nature of the contact forces is in progress. These results suggest that similar work on other brittle materials might be rewarding. Ice is not an ideal brittle material to work with, because it creeps significantly and is very close to its melting temperature.

Acknowledgements

This work was in part carried out within STRICE, a collaborative programme co-funded by the European Commission under the Energy, Environmental and Sustainable Development Programme of the 5th Framework Programme, under contract EVG1-CT-2000-00024. STRICE is led by HSVA (Hamburgische Schiffbau-Versuchsanstalt) and engages institutions in Finland, Sweden, Norway, France and the UK. Andrew Palmer's research is supported by the Jafar Foundation. He thanks John Dempsey for helpful conversations and Dev Sodhi for permission to use his tactile sensor data.

Bibliography

[1] Dempsey, J.P, Palmer, A.C. and Sodhi, D.S. High pressure zone formation during compressive ice failure. *Engineering Fracture Mechanics*, Vol. 68, pp. 1961-1974, 2001.

[2] Palmer, A.C. and Sanderson, T.J.O. Fractal crushing of ice and brittle solids. Proceedings of the Royal Society, London, series A, Vol. 433, pp. 469-477, 1991.

[3] Palmer, A.C. and Dempsey, J.P. Models of large-scale crushing and spalling related to high-pressure zones. Proceedings, IAHR Conference, Dunedin, NZ, 2002.

[4] Sanderson, T.J.O. Ice mechanics: risks to offshore structures. Graham & Trotman, London, 1988.

[5] Turcotte, D.L. Fractals and fragmentation. *Journal of Geophysical Research*, Vol. 91, pp. 1921-1926, 1986.

[6] Wadhams, P. Ice in the Ocean. Gordon and Breach Science Publishers, London, 2002.

[7] Xu, Y. Explanation of scaling phenomena based on fractal fragmentation. *Mechanics Research Communications*. In press (2004).

Fractal properties in rock fragmentation: result of a self similar process or consequence of a pure stochastic phenomenon?

Thierry Villemin[1] and Luc Empereur-Mot[1,2]

[1] LGCA, Université de Savoie, UMR CNRS 5025, 73376, Le Bourget du Lac, France

[2] Now at Dir. Syst. d'Info., Conseil Général de Savoie, 73018 Chambéry, France

Abstract

A numerical rock fragmentation model was elaborated, producing a 3D puzzle of convex polyhedra, geometrically described in a database. In the first scenario, a constant proportion of blocks are fragmented at each step of the process and leads to fractal distribution. In the second scenario, division affects one random block at each stage of the process, and produces a Weibull volume distribution law. Imposing a minimal distance between the fractures, the third scenario reveals a power-law. The inhibition of new fractures in the neighbourhood of existing discontinuities could be responsible for fractal properties in rock mass fragmentation.

1 Introduction

Natural and artificial fracturing lead to rock medium fragmentation, which finally evolves to a tridimensional puzzle of blocks. Fractures ending within the rock bulk are mechanically very unstable, and their growth make them meet other fractures, to finally produce a fully connected network. In both, natural [14, 17] and artificial fragmentation studies [3, 6, 7, 10, 14, 18], the number $N(r)$ of fragments of equivalent radius r (radius of the sphere having same volume as the block) satisfies the condition (1)

(1) $N(r) \sim r^{-D}$

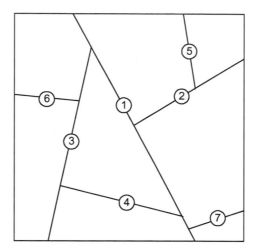

Figure 1: The block division process, involving successive fracture creation (1 to 7). In this example, fractures do not cross existing discontinuities.

This relation defines a fractal distribution of dimension D [19]. Trying to explain this law, several authors [1, 4, 17, 19] created or used a model whose main principle is as follows. Considering an initial cubic volume, divided in 8 identical cubic parts, each cube is given the same probability p to be divided in 8. This probability applies indefinitely to every resulting cube. By construction, this process leads to a distribution of N(r) that respects law (1). Moreover, D can be related to the fragmentation probability p by (2) [19]:

$$D = \ln(8p)/\ln(2)$$

But even if such a process gives a correct model of experimental fragmentation (1), the simplified geometry (puzzle of cubes) does not account for fragments shape and arrangement in real cases (e.g. [2, 12]).

2 The OBSIFRAC model

We developed a numerical model, named OBSIFRAC (*OBject SImulator for rock FRACturing*). From an initial, possibly pre-fragmented volume, plane fractures divide this volume into 2 or more son blocks. Figure 1 illustrate, on a 2D section, the first steps of the cutting process, which can be continued as far as needed. Such a process finally produces a fully-connected fracture network, in which every fracture abuts either on pre-existing fractures, or on the model boundaries. The fractured medium is therefore made up of a puzzle of polyhedral convex blocks, delimited by polygonal faces.

The OBSIFRAC model is connected to a relational database management system (DBMS), which undertakes fully geometrical description of the system, for every step of the process.

The program successively creates a set of i fractures $F_1, F_2, ..., F_i$. Starting from a *(i-1)* fractures network, the ith fracture is built after choosing its orientation (defined by its normal N_i), the seed point G_i from which it is extended, and the block Bi that it divides. The database is updated after each fracture creation, and keeps the memory of all successive states of the model. G_i, N_i, and B_i are chosen through different evolutionary motors.

3 Constant probability fragmentation

In this scenario, derived from Allègre et al. [1] model, a proportion p of blocks, chosen at random among previously generated blocks are simultaneously divided (salvoe). Point G and orientation N are chosen at random in each block. Unfragmented blocks are definitely excluded from the division process. The division probability *p* is the same for each block, and remains constant throughout the process.

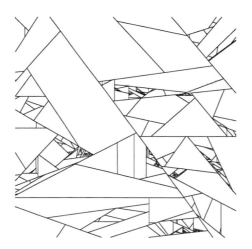

Figure 2: Constant-probability fragmentation with *p*=0,6. Initial state formed by N_0=101 blocks. Section through the model after 30 fracture salvoes (71841 blocks). Block size is extremely contrasted. Therefore, very few fractures are visible on this section. Smallest blocks gather in clusters.

On 2D sections, fracture traces gather in extremely dense clusters, revealing very contrasted block sizes (fig. 2 and 3). Blocks volumes have been studied at different stages of the process. Plotting number *N(R)* of elementary blocks of radius *R* (cubic root of volume) in log-log coordinates (fig. 3) shows a linearity, revealing a power-law distribution of the form of (1)

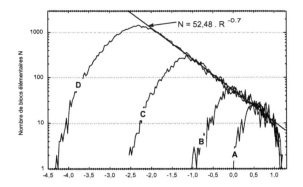

Figure 3: Constant-probability fragmentation with p=0,6. Initial state formed by N_0=101 blocks. Volume distribution after 5 fracture salvoes, 556 blocks (A), 10 salvoes, 1718 blocks (B), 20 salvoes, 11452 blocks (C), 30 salvoes, 71918 blocks (D). The useful portion of each curve shows a linearity, whose slope is constant and depends on the probability p.

(Fig. 3, Tab 1). The value for exponent D is about 0.7 for p=0.6. If p varies from 0.5 to 1, D values range from 0 to 3 (Tab. 1), and correspond to those given by relation (2). The main difference between Allègre *et al.* [1] and Turcotte [19] process, and our constant probability fragmentation process is that, in the latter, the volume of son blocks is not a constant fraction of the father block : the partition between the son blocks is not always the same, and depends on the position of the seed point G_i, and of the fracture orientation chosen in the father block. Our process is fractal, although point G_i is chosen random inside each block, and the volumes produced are random as well.

4 Random fragmentation

In this second scenario, one single block is chosen at random among the Ni blocks of step i, and will be cut. Each step produces one single fracture. The fragmentation probability (p=1/N_i at step i) obviously varies along the process and tends to 0.

p	0,5	0,6	0,7	0,8	1,0
D	0	0,7	1,5	2,1	3,0
D'	0	0,79	1,46	2,03	3,0

Table 1: Fractal dimension values, for constant-probability fragmentation, in Allègre *et al.* model [2] (D), and in OBSIFRAC model (D'). D and D' are quite similar for the studied values of the probability.

The 2D sections show a very heterogeneous fracture density, which is though less significant than in the constant probability case. Different distribution laws have been tested with Statistica ® V 5.1 software for volume distribution adjustment. $N(r)$ values do not fit lognormal, beta, gamma, power-law, Rayleigh, exponential, or extreme value laws. The maximum likelihood criterion method [13] reveals that the volume distribution closely matches a Weibull law (fig. 3b, $\Delta F=0$). The parameters of this law are then estimated by normal distribution transformation [9].

5 Random fragmentation with minimal distance between fractures

An existing discontinuity is likely to be preferentially reactivated, before any new fracture creation in its neighbourhood. The existence of a minimal distance Δ_F between fractures can therefore be conceived : at a shorter distance than Δ_F from each side of every fracture plane, fracture creation is mechanically very difficult. To take this feature into account in our random fragmentation model, we imposed that point G_i, which is chosen at random in the block, be situated at a greater distance than Δ_F from any other fracture delimiting the block.

Figure 4: Random fragmentation process, from one initial cubic block. Section in the model after 150000 fractures (150001 blocks). Block size is slightly less contrasted than in constant-probability fragmentation (Fig. 2). Fractures clusters are made up of the smallest blocks.

In the case of figure 5, a minimal distance Δ_F was fixed to 0.2 % of the total dimension of the model. Bigger values of Δ_F cause the simulation to stop prematurely, because of the impossibility to find any bloc matching the condition. On 2D sections, the fractured network is similar to figure 4. However, volume simulated data cannot be adjusted by a Weibull law in this case.

When Δ_F is given a positive value, and after elimination of extreme values related to finite-size effect in the model, the significant part of the distribution reveals a power law in the form of (1). For Δ_F=0,2 % of the size of the modelling volume, the exponent D is about 2.1. This value increases with Δ_F, and reaches 2.7 for Δ_F=0,1%.

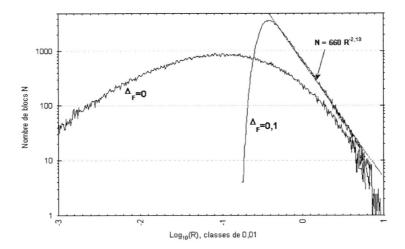

Figure 5: Random fragmentation process, from one initial cubic block. 150000 blocks. Volume distribution for a minimal distance between fractures Δ_F = 0 and 0,1.

6 Discussion and conclusions

For the 3 models presented above, where fractures orientations are random, similar results are obtained with more constrained orientations. Fracture orientation is therefore not an important factor for the *N(r)* law, and the case we presented is the most general one (random orientations).

Other protocols have been used to choose the values of (B_i, N_i, G_i). In particular, we demonstrated that the random choice of G_i in the whole modelling volume does not produce a distribution respecting relation (1). We restricted our presentation to processes involving power-law distributions.

On the basis of the simplified process imagined by Allègre *et al.* [1], and Turcotte [19], we tried to approach natural sections appearance (traces complexity, density). The orthogonal network used by these authors, just as models involving disk-shaped ([5, 15]) or rectangular fractures ([11]) cannot account for natural cases (e.g. [2, 12]). We demonstrated that, even if several random factors were introduced into the process (fracture orientation, fracture position in the affected block), the resulting volume distribution remained a power law, with the same exponent $D = \ln(8p)/\ln(2)$.

Although these 2 variation factors were added, the result does not give a realistic image of natural networks. Fractures gather in extremely dense clusters, distributed in the modelling volume. We imagined and tested a more realistic scenario, called random fragmentation, which clearly produces a Weibull law distribution for *N(r)*. Even though such a law approaches a fractal distribution for very small values of r, our result does not match experimental or field studies.

When a minimal distance between fractures Δ_F is imposed in our random fragmentation model, the *N(r)* distribution clearly approaches a power law, whose exponent increases with Δ_F. In this way, we demonstrate that adding such a limitation to an initial purely stochastic process leads to fractal properties. These results encourage us to think that the mechanical phenomenon inhibiting new fractures creation in the close neighbourhood of an existing fracture could be responsible for the existence of fractal properties in rock mass fragmentation. As clusters of small fragments are likely to weaken the whole blocks puzzle, this feature should be taken into account in mechanical models.

Bibliography

[1] Allègre C.J., Le Mouël J.L., Provost A., Scaling rules in rock fracture and possible implications for earthquake prediction. *Nature* 297 (1982) 47-49

[2] Barton C.C., La Pointe P.R., Fractal in the Earth sciences, Plenum Pr, NY, (1995) 265 p.

[3] Benett J.G., Broken coal. *J. Inst. Fuel* 10 (1936) 22-39.

[4] Biegel R.L., Sammis C.G., Dietrich J.H., The frictional properties of a simulated gouge having a fractal particle distribution. *J. Struct. Geol.* 11 (1989) 827-846.

[5] Billaux D., Hydrogéologie des milieux fracturés. Géométrie, connectivité et comportement hydraulique. Doc. BRGM 186 (1990), 277 p.

[6] Clark G.B., Principles of Rock Fragmentation. J. Wiley & sons, NY (1987) 610 pp.

[7] Curran D.R., Shockey D.A., Seaman L., Austin M., Mechanisms and models of cratering in earth media, in Impact and Explosion Cratering, D.J. Roddy, R.O. Pepin, R.P. Merrill, eds. pp. 1057-1087, Pergamon Press, NY (1977).

[8] Empereur-Mot L., La fragmentation naturelle des massifs rocheux. Modèles de blocs et bases de données tridimensionnelles. Géol. Alpine, Mém. HS n° 35, Univ. Grenoble (2001).

[9] Evans M., Hastings N., Peacock B., Statistical distributions. Wiley & sons, NY (1993).

[10] Fujiwara A, Kamimoto G., Tsukamoto A., Destruction of basaltic bodies by high-velocity impact. *Icarus* 31 (1977) 277-288.

[11] Gale J.E., Schaefer R.A., Carpenter A.B., Herbert A., Collection, analysis and integration of discrete fracture data from the Monterey formation for fractured reservoirs simulations, 66th Annu. Soc. Petrol. Eng. Tech.Conf., 22741 (1991), 823-834.

[12] Gillespie P.A., Howard C., Walsh J.J., Waterson J., Measurement & characterization of spatial distributions of fractures, *Tectonophysics*, 22 (1993), 113-141.

[13] Hahn G.J., Shapiro S.S.: Statistical models in engineering. Wiley & sons, NY (1967).

[14] Hartmann W.K., Terrestrial, lunar and interplanetary rock fragmentation. *Icarus* 10 (1969) 201-213.

[15] Long J.C.S., Gilmour P., Witherspoon P., A model for steady fluid flow in random 3D networks of disk-shaped fractures, *Water Resources Res.*, 21 (1985), 8, 105-115.

[16] Pascal C., Angelier J., Cacas M-C., Hancock P.L., Distribution of joints: probabilistic modelling and case study near Cardiff (Wales, UK). *Journal of Stru. Geol.*, 19, 10 (1997), pp 1273-1284.

[17] Sammis C.G., Biegel R.L., Fractals, fault gouge and friction. *Pure Appl. Geophys.* 131 (1989) 255-271.

[18] Schoutens J.E., Empirical analysis of nuclear and high-explosive cratering and ejecta, in Nuclear Geop. Sourcebook, 55, 2, 4, Rep. DNA OIH-4-2, Def. Nucl. Agency (1979).

[19] Turcotte D.L., Fractals and Chaos in Geology and Geophysics, 2nd Ed. Cambridge University Press (1997).

Application of the fractal fragmentation model to the fill of natural shear zones

David Mašín

Charles University, Institute of Hydrogeology, Engineering Geology and Applied Geophysics, Prague, Czech Republic
e-mail: masin@natur.cuni.cz

Abstract

Statistical descriptions of the frequency size distribution of rock fragments are summarised in the introduction. It is shown, that not all soils exhibit fractal frequency size distribution. A physical process leading to the scale invariant frequency size distribution of material sheared under constrained conditions is then described from the micro-mechanical point of view. The material from natural shear zones in silty shales is examined in the framework of this theory and it is shown that this theory may be relevant. Finally some possible applications for geotechnical engineering are discussed.

1 Introduction

A variety of statistical descriptions has been used to represent the frequency-size distribution of naturally and artificially fragmented material. An extensively used empirical description is the power-law scale-invariant (fractal) relation

$$N(> r) = Cr^{-D} \tag{1}$$

where $N(> r)$ is the number of fragments with a characteristic linear dimension greater than r, D is a fractal dimension and C is a constant chosen to fit the observed distributions. An alternative empirical correlation for the size-frequency distribution is the Weibull distribution given by

$$\frac{M(< r)}{M_0} = 1 - \exp\left[-\left(\frac{r}{r_0}\right)^v\right] \tag{2}$$

where $M(< r)$ is the cumulative mass of fragments with size less than r, M_0 is the total mass of fragments and r_0 is related to their mean size [12]. The Weibull distribution is entirely equivalent to the Rosin-Rammler [7] distribution, which is extensively used in geological applications. The Weibull distribution is not scale-invariant and reduces to fractal relationship only for small fragments.

Another example of an empirical relationship describing the fragmentation process is the log-normal distribution proposed by Epstein [5].

These examples, successfully applied in many experimental studies of the frequency-size distributions of rock fragments, show that not every crushed rock (soil) granulometry may be described using fractal relationship. Following the work of Sammis and Steacy [9], physical processes which should lead to scale-invariant distribution will be discussed theoretically in this paper and this hypothesis will be supported by examples of soil obtained from natural shear dislocations in silty shales.

2 The micromechanics of constrained comminution

A theory explaining scale-invariant frequency-size distribution of materials crushed under constrained conditions has been developed by Sammis et al. [8] in order to explain observed fractal distributions of rock fragments in crustal shear zones. This theory has been discussed and supported by laboratory observations in Sammis and Steacy [9].

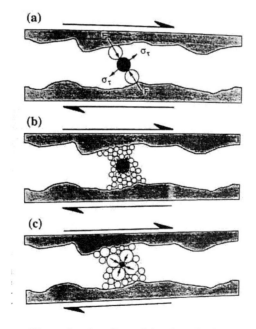

Figure 1: Schematic diagram illustrating the effect of the size of neighbouring particles on the tensile stress developed within a particle [9].

According to them frequency-size distribution is dependent on the way rock fragments are loaded. In industrial processes, such as pounding or tumbling, the volume of fragmented mass is unconstrained. Each fragment is equally likely to suffer impact, which means that the fracture probability of a given fragment is controlled by its strength. If fragment strength is independent of fragment size, then Epstein's [5] theory is relevant and a log-normal distribution is produced. If larger fragments are more fragile then the Rosin-Rammler [7] distribution is produced. In a fault zone, however, fragments are not free to change their relative positions without either additional fracturing or significant dilatational work. In this environment, a fragment's fracture probability of a fragment appears to be controlled primarily by the geometry of its neighbouring fragments that supply the load.

Consider the interaction of cylindrical particles illustrated in Fig. 1. In the case (a) the shaded fragment is being loaded by two fragments of the same size. Fragments loaded this way by a compressive force F fail by tensile splitting along the load axis. In the case (b), the shaded fragment is loaded by smaller neighbouring fragments. Although the net force along the stress path is the same, components orthogonal to the stress path reduce the tensile stress. The same is true for the case (c), in which the shaded fragment is loaded by larger neighbours. This hypothesis is supported by the experimental evidence illustrated in Figure 2 [10]. The larger the percentage of 3 mm beads in the mixture, the larger the probability they are in touch with another 3 mm bead and according to the theory the larger the probability they break.

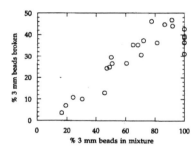

Figure 2: Percentage of 3 mm beads broken as a function of their fraction in mixtures of 3 and 1.5 mm beads [10].

If the fracture probability is maximum for neighbouring fragments of the same size (regardless their size), then the distribution evolves towards a geometry in which no two fragments of the same size neighbour at any scale. A simple fractal model, which has the same properties, is the array of cubes illustrated in Figure 3 [12]. In this model no two cubes of the same size are in contact (share face) at any scale except the smallest. The smallest cubes represent the lower fractal limit below which the power law is not valid. In the case of natural rock the lower fractal limit is delimited by the size of mineral grains. The fractal dimension D of this simple model is 2.58.

According to this theory, it is possible to explain fragmentation in the shear zone as follows: At the beginning, many large fragments have same-sized neighbours, but gradually large fragments become increasingly isolated. When there are no neighbouring large fragments (which are the most fragile due to the largest probability of

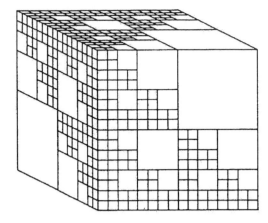

Figure 3: Fractal model for fragmentation [12].

flaws), the process shifts to the next largest size fragments for which many neighbouring pairs exist. Fragmentation process proceeds to ever smaller fragments until some fragmentation limit is reached (size of mineral grains). Then fragmentation begins again at the scale of largest fragments, since they are again the weakest, even though cushioned by smaller fragments. This continues with a cascade of fragmentation at progressively smaller scales.

The theory of constrained comminution therefore predicts fractal distribution of crushed material in the shear zone where the actual granulometry is dependent of the magnitude of tectonic movements.

3   Shear zones in the Mrázovka tunnel

The Mrázovka tunnel in Prague is designed as a part of the city highway system. The subsoil of the large diameter, two three-lane tunnels, consists mainly of clayey and silty shales of different age and weathering degree. The depth of the overburden varies from 15 to 40 meters. However, in the most critical stretch, under heavily developed urban environment the typical overburden is about 20 m. The geological site investigation revealed that tectonic shear zones intersect the tunnel profile at several locations [4]. These zones had been expected to be potential sources of unacceptable surface settlements during tunneling progress and therefore they were object of detailed investigation from the engineering-geological [3] and geotechnical [6, 1, 2] point of view.

These shear zones are usually from ten centimeters to several meters wide and they are filled with tectonically dislocated breccia-like material with varying granulometry depending on the degree of tectonical disintegration. Some shear zones are filled with smooth rock fragments "flowing" inside fine-grained matrix, in some cases the percentage of fine-grained material is smaller and the larger fragments create a skeleton. A typical small shear zone is shown in Figure 4 [3].

Figure 4: A photography of a small shear zone in the tunnel Mrázovka [3].

Mašín [6] studied the geotechnical characteristics of the fill of three shear zones — DPM1, DPM2 and DPM3 — in detail. Granulometric curves are shown in Figure 5, in which it is clear, that the proportion of fine-grained and coarse-grained material vary significantly.

According to the theory of constrained comminution, granulometry of the fill of such tectonic zones should obey fractal statistics and should be scale-invariant. The granulometric curves were re-plotted in the graph relating the number of particles larger than r as a function of their diameter r (Figure 6) using the following assumptions:

- The weight of the specimen is 10000 g, specific gravity of grains is 2.65.
- The particles have spherical shape.

It may be seen from Figure 6 that the power law relationship fits reasonably well the observed data and leads to fractal dimension 2.93. This is somewhat larger than the fractal dimension predicted by the model demonstrated in Figure 3 (D=2.58). This model is therefore too simple to describe fragmentation of the fill of tectonic shear zones. The results are in accordance with Turcotte [12], who points out that fractal dimensions of fragmented rock vary significantly, but most lie in the range $2 < D < 3$.

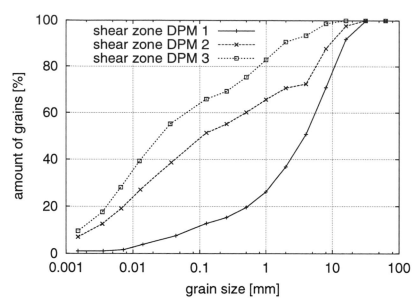

Figure 5: Granulometry of the fill from three typical shear zones in silty shales from the Mrázovka tunnel [6].

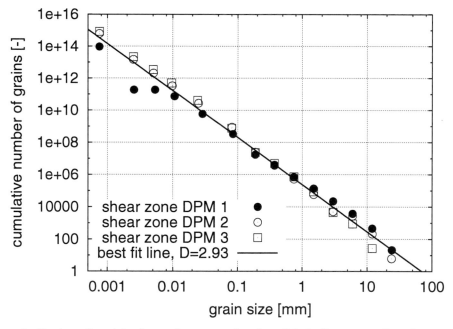

Figure 6: Number of particles larger than r as a function of their diameter r. Best fit power law relationship leads to fractal dimension D=2.93.

The constrained comminution theory can describe the variability of the grain size distribution of materials from different tectonic zones. Different degrees of tectonic dislocation leads to different positions in the cascade of fragmentation described in the previous section.

4 Discussion of possible applications

In recent years the finite element method became a standard tool in predicting the behaviour of various geotechnical structures. In order to perform reliable analysis, it is necessary to use constitutive models which can describe with sufficient accuracy soil behaviour. Many advanced constitutive models have been developed in past years and they have been successfully applied for analyses of different geotechnical problems. Most advanced constitutive models restrict the range of their validity either to fine grained or coarse grained soils, the behaviour of their mixtures is however not fully understood. This fact is limiting, as most natural soils have gradual granulometry curves rather than being pure sands or clays.

During the investigation of the geotechnical properties of the fill of tectonic shear zones in the Mrázovka tunnel a simple theory for the behaviour of soil mixtures proposed by Thevanayagam and Mohan [11] was applied by Mašín [6] . According to this theory, it is possible to predict the behaviour of the soil mixture only on the basis of the knowledge of the behaviour of fine grained component and coarse grained component and the amount of fine grained component in the mixture. The soil should behave as a pure coarse grained component if the void ratio of coarse grained fraction (e_g) is smaller than the maximum void ratio, which the coarse grained skeleton can sustain (e_g, max). On the other hand, if grains of the coarse grained fraction "float" in the fine grained matrix and do not create a skeleton, then the behaviour is governed by the fine-grained fraction. The range between these two extremes is called "transition zone". Experimentally determined dependence of the peak and critical state friction angles on the amount of fine-grained fraction in the mixture, showing this transition zone, is shown in Figure 7 [6].

A shortcoming of this theory is that the granulometry curves of natural soils are gradual and contain fractions of all sizes, therefore it is not possible to clearly distinguish between fine grained and coarse grained fractions. The second problem, which comes up, is that the size of the largest particles in the mixture is often too large and the material can not be tested in the standard laboratory equipment. Researchers are therefore forced to perform laboratory tests on material with reduced granulometry curve (with the largest particles removed). Application of results from these tests to the original material is however problematic.

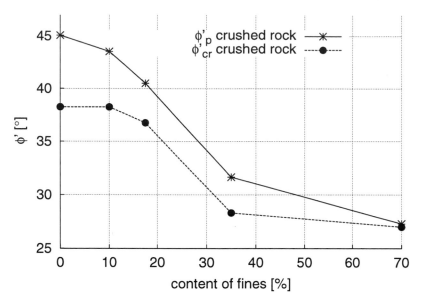

Figure 7: Dependence of the peak and critical state friction angle on the amount of fi ne-grained fraction in the mixture [6].

The fact that the granulometry of (at least some) soils can be with sufficient accuracy described by fractal, scale-invariant laws could be possibly used to construct more advanced conceptual model for soil mixtures and to help understanding their mechanical behaviour. Also, fractal distribution of grain sizes in the mixture could help in evaluation of laboratory tests on materials with reduced granulometry curve.

5 Conclusions

The fill material from three tectonic shear zones in silty-clayey shales has been investigated in the framework of the theory of constrained comminution, proposed by Sammis and Steacy [9]. It has been shown that this material exhibits fractal, scale-invariant distribution, and this theory is a possible micromechanical explanation of the process leading to the observed grain size distribution. Some possible applications of these observations for soils with gradual granulometric curves have been discussed.

Acknowledgement

The author would like to thank Prof. D. Kolymbas for valuable suggestions. Financial support by the research grant GAAV A2111301 is also gratefully acknowledged.

Bibliography

[1] J. Boháč, I. Herle, and D. Mašín. Stress and strain dependent stiffness in a numerical model of a tunnel. In *Proc. 2nd Int. Conference on Soil Structure Interaction in Urban Civil Engineering, Zürich*, 2002.

[2] J. Boháč and D. Mašín. Laboratory modelling of soil from faults in silty shales (in Czech). In *Proc. 5th Int. Conference Geotechnical structures optimization, Bratislava*, 2001.

[3] R. Chmelař. *Characterisation of the rock massive when NATM is applied in a dislocated zones of tunnel Mrázovka (in Czech, in preparation)*. PhD thesis, Charles University, Prague, 2003.

[4] R. Chmelař and J Vorel. The problematic of geotechnical survey of the rock massive during NATM excavation of tunnel Mrázovka (in Czech). In *In proc. Podzemní stavby*. Prague, 2000.

[5] B. Epstein. The mathematical description of certain breakage mechanisms leading to the logarithmico–normal ditribution. *J. Franklin. Inst.*, 224:471–477, 1947.

[6] D. Mašín. The influence of filling of tectonic joints on tunnel deformations (in Czech). Master's thesis, Charles University, Prague, 2001.

[7] P. Rosin and E. Rammler. Laws governing the fineness of powdered coal. *J. Inst. Fuel*, 7:89–105, 1933.

[8] C. Sammis, G. King, and R. Biegel. The kinematics of gouge deformation. *Pure Appl. Geophys.*, 125:777–812, 1987.

[9] C. G. Sammis and S. J. Steacy. Fractal fragmentation in crustal shear zones. In C. C. Barton and R. La Pointe, P, editors, *Fractals in the Earth Sciences*, pages 179–204. Plenum Press, New York, 1995.

[10] S. J. Steacy. *The Mechanics of Failure and Fragmentation in Fault Zones*. PhD thesis, University of Southern California, 1992.

[11] S. Thevanayagam and S. Mohan. Intergranular state variables and stress–strain behaviour of silty sands. *Géotechnique*, 50(1):1–23, 2000.

[12] D. L. Turcotte. *Fractals and Chaos in geology and geophysics*. Cambridge University Press, Cambridge, 1992.

The Development of Fractal Geometries in Deformed Rocks

Bruce E. Hobbs and Alison Ord

CSIRO Exploration and Mining, Perth, Australia

email: bruce.hobbs@csiro.au and alison.ord@csiro.au

Abstract

Structures in deformed rocks are scale invariant: similar geometry exists over length scales from tens of kilometres to the millimetre scale. Such scale invariance is the hallmark of fractal geometry. This paper illustrates scale invariance with reference to the buckling of multi-layers. Simple elastic-viscous constitutive relations lead to periodic fold systems with no scale invariance. Combinations of these constitutive laws can produce localised folds with irregularity and fractal geometry. Constitutive laws involving microstructure comprised of alternating layers of anisotropic viscous and less viscous material produce fold systems with scale invariance via a parameter involving the fineness of the layering. Here, mode interaction also produces folds at two length scales. It is not clear that this represents a second bifurcation in the system, however a cascading sequence of bifurcations could be another mechanism for producing scale invariance.

1 Scale Invariance, Fractals and Emergence: Some Definitions

In order to clarify the discussion below, we begin by indicating the ways in which we use a number of terms in this paper. First, the term *scale invariance* means that in a particular geometrical system, similar types of structures are seen over a wide range of length scales. The term does not imply that the system is self-similar. *Self-similar* means that the geometry at one length scale may be derived from that at another length scale by means of a deformation that involves only a homogeneous dilation (see Figure 1). The term deformation is used to mean a geometrical transformation of points. Self-similar is quite different to the term *self-affine*, which means that the geometry at one length scale may be derived from that at another length scale by means of a general homogeneous deformation that may involve shear as well as dilation (see [1]). A classical case of self-affine geometry in geomaterials is the surface of a fault or joint plane. Here, the geometry at a large enough scale is planar and, for all intents, Euclidean. However at finer length scales roughness appears. If the geometry of a joint is self-affine then these surface roughnesses at fine scales can be derived from the much longer length scale irregularities observed at the large length scale by a homogeneous shortening parallel to the mean orientation of the surface. The distinction is important because different measuring procedures are required for self-similar and self-affine geometries (see [4]) and misguided statements can be made if this distinction is not recognised. Presumably there exist a class of geometries that are *self-non-affine*; that is the geometry at various scales can be derived from that at other scales by a general *inhomogeneous* deformation. To our knowledge no one has ever investigated such geometries but it is distinctly possible that many geomaterials are characterised by this class of geometry (see Figure 1).

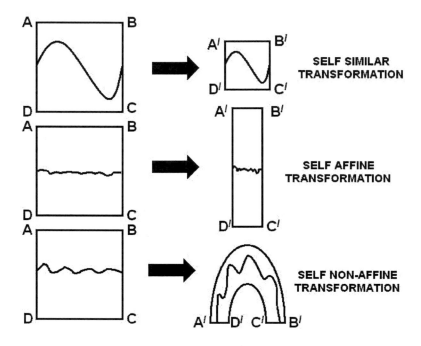

Figure 1: Examples of self-similar, self-affine and self-non-affine transformations.

There then arises the question of whether the types of geometries referred to above are fractal. The term *fractal* is used here in the sense of Feder [3] to mean *a system made of parts similar to the whole in some way*. This means that, by definition, scale invariant systems are fractal. An equally relevant quality of fractal systems is that proposed by Mandelbrot [8, p 361]: *fractal structures are invariant under some suitable collection of smooth transformations*. However, there is an almost universal expectation that the term fractal carries with it the additional implication that the system is characterised by a unique non-linear, non-integer length scaling law that enables the geometry to be scaled from one size to another. We do not intend that this implication is relevant to the types of fractals that are geomaterials. What we do intend is that if one were to construct the attractor for the geometry in the manner presented by Ord [11] by analogy with the long established method for characterising time-series (see [14]) then that attractor would be "strange" or "wild" in the sense that it would possess a non-integer scaling law. To us then, the term *fractal*, as applied to geological systems, means a scale invariant geometry characterised by a strange attractor. Unfortunately, the construction of an attractor for a geomaterial system depends on the ability to collect accurate three-dimensional geometric data and this, to date, has been difficult. The development of various laser scanning and photogrammetric techniques in recent years now holds promise of rapid developments in this field (see [12] and Ord, this volume).

Finally, there is also an implication that fractal geometries arise from feedback processes between those non-linear mechanisms that were responsible for the development of the geometry. Thus the field of non-linear mechanics is commonly linked to the formation of fractals. This is the topic of the present paper, and we will use the buckling of a layer embedded in other materials as an example. The topic concerns the concept of *emergence,* which is a property of large dissipative systems driven far from equilibrium (see [7]). The term refers to the spontaneous development of structure or patterning in a system that previously had been homogeneous. This can be thought of as a phase transition [17]. The development of fracture systems, shear zones, fold systems, zoned mineral alteration systems (including ore bodies), and fluid percolation networks can all be thought of as emergent phenomena in geomaterials.

Emergent behaviour is associated with *bifurcation* in the system of differential equations that describes the mechanics of the system. A classical simple example of bifurcation (see [7], for a detailed discussion) is the behaviour of the solutions to the equation:

$$d\alpha/dt = -\alpha^3 + \lambda\alpha \qquad (1)$$

where λ is parameter such as amount of strain. When $\lambda < 0$ there is only one real stationary state, $\alpha = 0$; when $\lambda > 0$ two real stationary states exist, $\alpha = \pm\sqrt{\lambda}$. $\lambda = 0$ is a bifurcation point (see Figure 2a). The typical stress strain curve for a geomaterial is an example of bifurcation behaviour (see Figure 2b). Here the material behaviour departs from homogeneous deformation to localised deformation at a bifurcation point determined by the amount of strain.

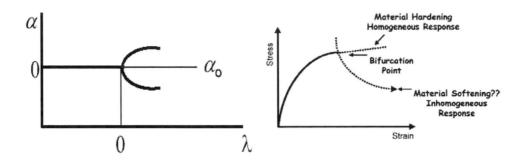

Figure 2: Bifurcation behaviour. (a) Behaviour of equation (1). (b) A typical stress strain curve showing the departure from homogeneous to localised deformation at a bifurcation point determined by
the amount of strain.

The emphasis in this paper, then, is on the processes that lead to scale invariance
in the geometry of deformed rock masses. The motivation is that if we could understand these processes and their controls on the emergent geometry we would be
in a better position to make predictions about the unexposed geometry in rock
masses and develop robust, physics based methods of interpolation and extrapolation to aid mineral exploration and mining/civil construction design.

2 Folding of Layered Rocks

Folds in layered geological materials occur at all scales ranging from the 10's of
kilometer scale down to the millimetre scale. Moreover the geometrical character,
or style, of these folds is identical at all scales so that they appear to comprise a
scale invariant set. The ratio of wavelength to layer thickness of such natural fold
systems is typically of the order of 4 to 7 whilst the calculated contrast in mechanical properties (e.g. viscosity) between adjacent layers, based on laboratory
experimental results, is typically of the order of 10 to 40 [15]. At a particular scale,
the profiles of the fold system are never strictly periodic; many wavelengths are
present at the particular scale under consideration and the fold train as a whole
tends to be better described as irregular rather than periodic (see Figure 3).

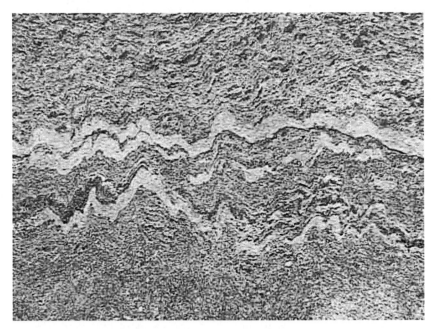

Figure 3: Irregular fold system in gneiss.

3 Folding Theory

These observed geometrical characteristics contrast with what is predicted from relatively simple analyses of layered isotropic elastic or viscous materials such as those developed by Biot [2] and Ramberg [16]. These analyses, at least for a single layer embedded in a less viscous matrix or in a matrix with substantially lower elastic modulus, predict strictly periodic wave forms with only one wavelength (the so called dominant wavelength) amplified. The wavelength that is preferentially amplified depends upon the contrast in mechanical properties (elastic modulus and/or viscosity) between the layer and the embedding matrix. Significant amplification occurs only if this contrast is of the order of 100 or more. For lower contrasts in mechanical properties, homogeneous layer shortening occurs to high strains where explosive amplification of periodic fold trains may ensue. For the range of mechanical contrasts where folds readily amplify, the ratio of wavelength to layer thickness is typically 10 to 25. Thus it is clear that single layer analyses of materials with relatively simple constitutive behaviour (isotropic elastic or viscous) do not match the observations that can be made on natural fold systems.

Moreover, the situation is not improved by examining the behaviour of multilayered materials that have constitutive behaviour that is isotropic and elastic or viscous. Again, only one wavelength is amplified and this wavelength tends to be controlled by the behaviour of the thickest layer. The remaining thinner layers are distorted in a more or less passive manner to conform to the behaviour of the thickest layer.

It is important to note that in all of the above examples, linear stability analyses indicate just one bifurcation point at the onset of buckling and numerical simulation shows that no further bifurcation points are met as the deformation proceeds; the initial instability is simply amplified by continued deformation.

As is to be expected, the introduction of more complicated constitutive behaviour introduces greater complexity into the resulting geometry. Thus isotropic elastic-viscous materials do produce wave trains that, in profile, are not periodic and under some circumstances, two distinct wavelengths are amplified [10]. However the wavelengths that are amplified are commonly close (but different to) the Biot wavelength and so the resultant wavelength to thickness ratios are still too high to be representative of natural fold systems.

Even more complicated isotropic elastic-viscous materials [6] lead to the development of localised fold trains so that periodicity in the fold profile is now lost. The development of these localised fold packets is strongly dependent on initial conditions (displacement and velocity) and for some initial conditions distinct fractal behaviour ensues.

Thus the incorporation of elasticity, plasticity and viscosity in isotropic constitutive relations captures the essence of the problem: First, the dominant wavelength is dependent on layer thickness and is a function of the contrast in mechanical properties. Second, for materials with non-associated plasticity, shear banding is geometrically related to folds in naturally observed ways. Third, for some exotic constitutive relations localisation of fold packets occurs and the geometry becomes irregular; some fractal characteristics emerge but scale invariance, if present, extends over only a small range of length scales. The important issue is that these isotropic constitutive laws lead to wavelength to thickness ratios that are too large compared to natural observations and require contrasts in mechanical properties that are too large to be compatible with experimentally determined mechanical properties.

4 Behaviour of Anisotropic, Micro-Layered Materials

However the real approach to natural fold geometries arises once anisotropy is introduced into the constitutive behaviour, be it elastic, viscous, plastic or some combination of these three material behaviours. This anisotropy needs to be incorporated via a fine scale layering, parallel to the bulk layering, comprised of alternating contrasting anisotropic, viscous material [5], [9]. Once such anisotropy is introduced, the folding instability is driven by contrasts in the layer parallel mechanical property (for instance the layer parallel shear viscosity) rather than the layer normal mechanical property [5].

If we consider the anisotropic layer to be comprised of alternations of viscous and less viscous fine layers with each pair of viscous/less viscous layers comprising a unit of thickness h, then, taking the dispersion relation from Hobbs et al. ([5], equation 19), we derive the dominant wavelength, λ, for anisotropic finely layered materials as:

$$\lambda = 2\pi H \{ \beta \, \Phi \, \Delta\eta \, / \, 6\eta_e \}^{1/3} \tag{2}$$

where H is the layer thickness, $\Delta\eta$ is the difference in the bulk layer normal viscosity, η, and the bulk layer parallel, shear viscosity, η_s that is $\Delta\eta = \eta - \eta_s$, and η_e is the viscosity of the embedding medium. β is the ratio, h/H, of the thickness of an anisotropic unit and the total layer thickness; β in natural rocks is quite variable with a range from 10^{-1} to 10^{-6}. Φ is a geometrical factor that involves the volume fractions of viscous and less viscous fine layers and their relative viscosities [5]; a reasonable range in values is 0.01 to 0.75 for the case $\eta_s = 0.1\eta$.

Expression (2) is to be compared with the classical Biot expression for the dominant wavelength:

$$\lambda = 2\pi H \{\eta \, / \, 6\eta_e \}^{1/3} \tag{3}$$

where η now is the viscosity of the (isotropic) embedded layer. For an anisotropic material where η_s is much less than η, so that $\Delta\eta \approx \eta$, the difference between expressions (2) and (3) is approximately $\{\beta \, \Phi\}^{1/3}$. If, for instance there are equal proportions of viscous and less viscous fine layers, $\Phi = 0.25$ (see discussion in [5]), and if we take $\beta = 0.001$, then for an anisotropic, layered viscous material,

$$\lambda = 2\pi(6.3 \times 10^{-2})H \{ \Delta\eta \, / \, 6\eta_e \}^{1/3}$$

so that the wavelengths predicted here are very fine and clearly much smaller than typical Biot type wavelengths. A reasonable range for $\{\beta \, \Phi\}^{1/3}$, assuming $\eta_s = 0.1\eta$, is approximately 0.5 to 2×10^{-3} which is equivalent to about two orders of magnitude in length scales.

In addition, the anisotropic viscous theory predicts amplification at much smaller viscosity contrasts, $(\eta \, / \, \eta_e)$, than in the classical Biot theory. For instance, for anisotropic viscous materials, folds with ratios of wavelength to thickness of 5 are amplified for contrasts in layer normal viscosity of 20, and layer parallel shear viscosity of 40 (see Figure 4). If this situation is modelled with both layers isotropic and the viscosity contrast 20, no folding occurs up to a shortening of 50%. Moreover, anisotropic materials appear to undergo a sequence of bifurcations following the initial bifurcation corresponding to the first folding instability. Linear stability analysis predicts the presence of the first instability but numerical simulation shows that continued deformation introduces additional folding instabilities. Thus, in Figure 4, fine scale folds develop very early in the deformation as predicted by the linear stability analysis. These produce quasi-periodic irregularities on the top and bottom of the layer through some kind of poorly understood long-range interactions between the initial buckles. This represents some form of mode interaction [13], [9]. These irregularities are not only homogeneously amplified as the deformation continues but grow in their own right. This represents a second bifurcation (a so-called *svegliable* or "capable of being awakened" instability in the sense of Ortoleva [13]). It seems that some of the observed scale invariance of natural fold systems arises from this multiple bifurcation sequence.

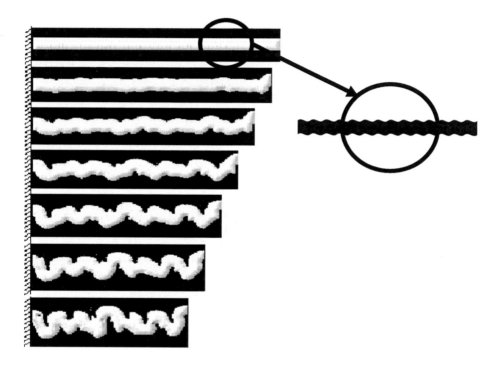

Figure 4: Progressive development of a fold system in anisotropic, finely layered, viscous material. The central layer is itself layered on a fine scale (after [5]). The inset shows a representation of the fine scale folds that develop initially in the bulk layer.

5 Summary

Perhaps there are three (closely integrated) processes involved in the generation of scale invariant geometries.

1. The first process involves the intrinsic presence of scale invariant geometry to start with. For instance, anisotropy, together with fine scale layering, seem to be fundamental in producing naturally observed geometries. The presence of layering at a large range of scales is ubiquitous in natural rock sequences. The calculated range in $\{\beta\ \Phi\}^{1/3}$ of two orders of magnitude implies a similar range in scale in the wavelengths of fine scale folding.

2. Non-classical, non-linear constitutive relations, such as those considered by Hunt and co-workers [6], contribute to the development of fractal behaviour and irregular fold geometries. Their contribution to a wide range in scale invariant length scales is not clear however.

3. Instabilities arising from svegliable [13] behaviour are responsible for another class of fold wavelength scales approximately an order of magnitude greater than those arising from item 1 above. This process may involve mode interaction and/or a new bifurcation mode. Either way, the process is capable of producing a new range of fold wavelengths intimately associated with initial fine scale layer thickness distributions. A cascading sequence of bifurcations (see Figure 5), each associated with a particular wavelength is a particularly powerful and elegant way of developing scale invariance.

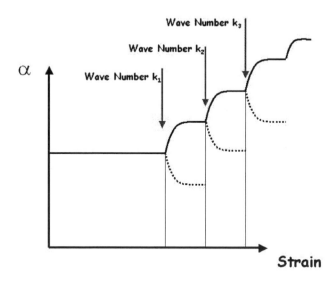

Figure 5: Cascade of bifurcation points with increasing strain, each associated with a new wavelength.

Bibliography

[1] Barnsley, M, F.: Fractals Everywhere. Academic Press, New York, 396 pp, 1988.

[2] Biot, M. A.: Theory of folding of stratified viscoelastic media and its implications intectonics and orogenesis. *Geol. Soc. Am. Bull.* 72, 1595-1620. 1961.

[3] Feder, J.: Fractals. Pleneum Press, New York, 283pp, 1998.

[4] Hobbs, B. E.: The significance of structural geology in rock mechanics. John. A. Hudson (Ed), Comprehensive Rock Mechanics, Pergamon Press, Chapter 2, 25-62, 1993.

[5] Hobbs, B.E., Muhlhaus, H-B, Ord, A and Moresi, L.: The Influence of Chemical migration upon Fold Evolution in Multi-layered Materials. Vol. 11, Yearbook of Self Organisation. Eds H.J. Krug and J.H. Kruhl. 2001.

[6] Hunt, G.W. Wadee, M.K. & Shiacolas, N.: Localized elasticae for the strut on the linear foundation. *J. Appl. Mech.* 60, 1033-1038. 1993.

[7] Kondepudi, D. and Prigogine, I.: Modern Thermodynamics: From Heat Engines to Dissipative Structures, John Wiley and Sons, Chichester. 1998.

[8] Mandelbrot, B, B.: The Fractal Geometry of Nature, Freeman, San Francisco, 468pp, 1982.

[9] Muhlhaus, H-B, Moresi, L. N, Hobbs, B. E, and Dufour, F.: Large Amplitude Folding in Finely Layered, Viscoelastic Rock Structures. *Pure and Applied Geophysics,* 159, 2311-2333, 2002.

[10] Mühlhaus, H.-B. Sakaguchi, H. & Hobbs, B.E.: Evolution of three-dimensional folds for a non-Newtonian plate in a viscous medium. *Proc. R. Soc. Lond.* A 454, 3121-3143. 1998.

[11] Ord, A.: The fractal geometry of patterned structures in numerical models for rock deformation. In: J.H. Kruhl, Fractals and Dynamic Systems in Geoscience, Springer-Verlag, Berlin, 131-155, 1994.

[12] Ord, A, and Cheung, C, C.: Image analysis techniques for determining the fractal dimensions of rock joint and fragment size distributions. Computer Methods and Advances in Geomechanics, Beer, Booker and Carter (Eds) Balkema, Rotterdam, 87-91, 1991.

[13] Ortoleva, P. J.: Geochemical Self-Organisation. Oxford University Press. 411 pp. 1994.

[14] Packard, Crutchfield, Farmer, and Shaw.: Geometry from a time series. *Phys. Rev. Let.,* 45, 712, 1980.

[15] Patton, R.L and Watkinson, A.J.: A viscoelastic strain energy principle expressed in fold-thrust belts and other compressional regimes. *J. Struct. Geol.* In press. 2004.

[16] Ramberg, H.: Fluid dynamics of viscous buckling applicable to folding of layered rocks. *Bull. Am. Assoc. Pet. Geol.* 47, 484-505. 1963.

[17] Sethna, J.P.: Order parameters, broken symmetry, and topology. In: Lectures in the Sciences of Complexity. Ed. L. Nadel and D. Stein. 243-288, Addison-Wesley, New York. 1992.

3D Imaging of Jointed Rock Masses

Alison Ord[1], Fabio Boschetti[2], Bruce E. Hobbs[3]

CSIRO Exploration and Mining, Predictive Mineral Discovery Cooperative Research Centre, Perth, Australia

[1] e-mail: alison.ord@csiro.au

[2] e-mail: fabio.boschetti@csiro.au

[3] e-mail: bruce.hobbs@csiro.au

Abstract

With the advent of high-resolution digital cameras it is now possible to collect, remotely, three-dimensional data on the geometry of rock joints. This means that for the first time we can explore if such geometry is truly fractal. In this paper we explore three techniques (iterative function systems, phase space reconstructions and Epsilon Machines) for (i) exploring three dimensional joint geometry in an attempt to place constraints on the fundamental physical processes responsible for joint formation and (ii) for providing better extrapolation and interpolation procedures for making predictions of rock joint geometry for geotechnical, mine design and mineral exploration studies.

1 Introduction

In order to improve our ability to make predictions for geotechnical, mine design and mineral exploration studies it is important that we develop better ways, ideally based on a fundamental physical understanding of the underlying processes, of extrapolating, interpolating and scaling the geometry of rock joints based on necessarily limited field observations. Hence it is fundamental that we understand if the geometry of rock joints is truly fractal or something else. There have been a large number of studies of the geometry of rock joints [13, 15], and log-normal [20], fractal [4, 3] and Weibull [1] distributions have been reported. Unfortunately, and by necessity, all of the available studies of rock joint geometry are in one or two dimensions. Here we report on ways of establishing three-dimensional geometry and some possible ways of analysing and interpreting the data.

The purpose of this paper is therefore to review methods for collecting data on the three dimensional geometries of rock joints, and to illustrate some new advances in analysing these data using the concepts of fractal geometry. In particular, we describe three different approaches to analysing these data using functional (iterated function systems), geometrical (phase space reconstructions) and algorithmic (Epsilon machines) approaches.

Rock masses are commonly broken up into a number of discrete blocks by two or more joint sets. The joints themselves are a result of brittle behaviour during regional deformation, of differential volume change in heterogeneous bodies, and/or of erosional unloading [11]. Examples here are taken from Central Australia and the Ok Tedi mine in Papua New Guinea.

In Central Australia, a well sorted, silicified and metamorphosed quartzite, the Heavitree Quartzite, forms the core of the Macdonnell Ranges. This rock unit is notable for its dominance of the skyline, its horizontal extent (~500 km east-west), and its remarkably persistent joint pattern - three almost orthogonal joint sets. The joint sets are observed at scales from kilometres (Figure 1a) down to metres (Figures 1b, 1c, 1d) and to centimetres (Figures 1e, 1f). Another example involves three almost orthogonal joint sets seen in the mine access road at Ok Tedi, Papua New Guinea (Figure 2).

1a

1b

1c

1d

Figure 1: Joint sets of various scales

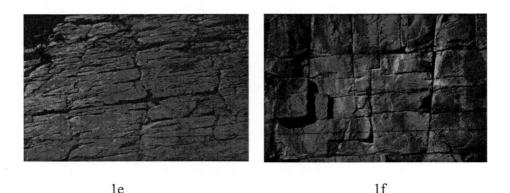

1e   1f

Figure 1: Joint sets of various scales

Figure 2: Joint sets in mine wall at Ok Tedi mine, Papua New Guinea.

The fundamental issue is whether or not we may infer anything from these joint structures about the mechanics of their formation. The practical issue is being able to predict the spacing, persistence, and orientation of rock joints sufficiently well to provide physically based, geologically realistic 3D models for, at increasing scale, mining processes, mine stability, and mineralisation, involving deformation, fluid flow, thermal transport, and chemical reaction. Approaches to this problem, generally statistical, have become increasingly complex over the years. However, the predicted patterns lack naturalness; they miss some essential geological truth [9]. It is our thesis that what the predicted patterns lack is a representation of the underlying physical and geological processes of formation for the rock joints. However, the relationships/constitutive behaviours which do exist to describe processes of rock jointing are not yet able to predict the observed characteristics of 3-dimensional rock joint patterns. The question addressed in this paper is how may we begin to fill such a large gap between theory and reality?

2 A Step Forward

A second major issue for a useful analysis of such joint sets is the small amount of data from which such analyses are usually made. Classically [12], a metre square of the rock face would be measured, and the joints within that square metre described with respect to their orientation, spacing, and persistence. Such a square metre might be measured every 50 to 100 m along a mine face. Unstable conditions of the wall can mean that such measurements would be estimated at some distance from the wall, or just not determined at all. Such sampling is, therefore, hardly representative. Above all, although data on orientation, spacing and persistence are commonly collected, no three dimensional spatial data in terms of x, y, z coordinates are normally collected.

Remote imaging techniques for the mine face topography were early perceived to provide an opportunity for obtaining such information better, faster, and more safely; so much more data may be obtained that a step change should occur in our understanding and prediction of rock joint patterns. Over the past two decades, both active and passive imaging techniques have been explored [5, 17] for this purpose. In active imaging, a structured light source illuminates the scene, which on a flat plane surface would reflect a regular and known pattern. The perturbations to this pattern caused by intersection of the light with an irregular surface are captured by a camera (Figure 3a). The three dimensional geometry of the surface causing these perturbations may then be calculated from the perturbed reflected light image. In this instance, a laser is used to project a planar beam across the rock face (Figure 3b). The perturbations arising from the intersection of the beam with the topography of the rock face is collected by the camera. The computer calculates (x, y, z) based on the degree of perturbation from a flat plane (Figure 3c). Passive imaging, involving photogrammetry employing 'normal' photography is presently favoured as a result of improvements in hardware and software [19, 23] with an image of a reconstructed rock face shown in Figure 4.

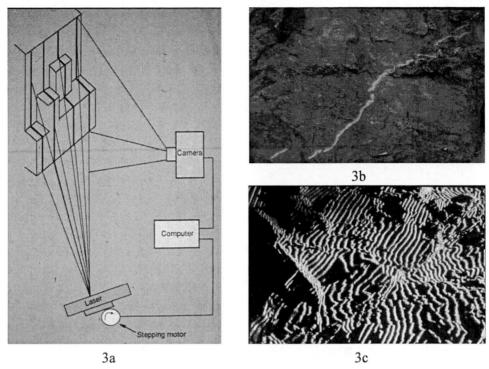

3a   3b
3c

Figure 3 SIROJOINT: active approach. (a) Sketch of the concept. (b) Intersection of laser beam with rock face. (c) Repeated intersections of laser beam with rock face, from which the topography may be reconstructed.

Figure 4: Reconstructed rock face

3 Conceptual Explorations

We describe here three conceptual approaches which we have begun to explore in our aim to determine the underlying physics of the system. These approaches are iterated function systems [2], phase space reconstructions [18], and epsilon machines [7].

3.1 Iterated function systems

Iterated function systems are an approach to describing a complex structure through a small set of simple (affine) geometrical transformations. Barnsley showed that a set of affine transformations, when applied randomly to some initially randomly chosen point, produces a complex geometry and, with care in choosing the transformations, can produce as natural a shape as a fern [2].

We have developed our own software to explore the consequences of this approach for geological patterns, particularly joints and fractures, and folds [9, 10]. So for example the set of five affine transformations represented by the parallelograms shown in Figure 5a, or just 20 numbers, when applied to one point, results in the pattern of points shown in Figure 5b. Patterns such as this are strongly reminiscent of patterns of ore grade distributions such as gold and copper at Ok Tedi in Papua New Guinea, and vein geometries. Figure 5c shows the same pattern of points, but contoured by spatial density of the points. So this joint set or ore grade distribution may be represented for geotechnical and mine design purposes by just 20 numbers - this must surely be the most efficient quantitative way of describing natural structures, and the approach can readily be extended to three dimensions for incorporation of joint generation routines into mine stability software.

3.2 Phase space reconstructions

This second approach [18, 24, 6] is based on Poincare's work of the late 1800's. Delayed coordinate embedding involves discretisation of the system through the use of increasing multiples of a fixed space lag. Through this approach, the system's dynamics are 'unfolded' into a multidimensional phase space. If the trajectories within this phase space converge to a subspace or geometrical attractor, may then the embedding dimension of this attractor indeed represent the underlying dynamics of the system? The traditional use of this approach is for describing temporally-patterned phenomena. We have explored this approach [16] for a spatially-disposed pattern, specifically a computationally produced series of shear bands, using an explicit finite difference scheme for deformation [14].

We attempted at this time to apply the same approach to 3D data obtained on folded rocks using an early prototype of SIROJOINT. However, there was still not enough data and, as demonstrated further in the next section, there was also the problem of the lack of data in various areas as a result of the irregularities in the rock surface, and shadowing of certain areas.

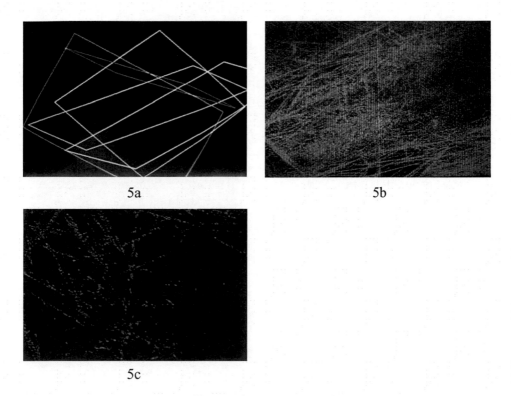

5a

5b

5c

Figure 5: Affine transformations and their results

3.3 Epsilon machines

Epsilon machines are mathematical expressions of the 'computational mechanics' developed by James Crutchfield and co-workers at the Santa Fe Institute, New Mexico [7].

Their thesis is that one must consider pattern discovery, not pattern recognition, and not building predictors, neither of which will lead to any genuine understanding of the process. Pattern discovery on the other hand asks the question: given the data produced by a process, how can one extract meaningful, predictive patterns from it, without fixing in advance the kind of patterns one looks for or the forward modelling algorithm one will use to reproduce the pattern? This is precisely what we wish to know in many geological and geotechnical enquires. The problem is exemplified by a quote ([21, from Plato, Meno]:

'How will you look for it, Socrates, when you do not know at all what it is? How will you aim to search for something you do not know at all? If you should meet it, how will you know that this is the thing that you did not know?'

Crutchfield describes the problem as: Given a dynamical system, how can one detect what the system itself is intrinsically computing?

The reconstruction of an epsilon machine works by recognising the 'states' which the system visits at different times, and the transition rules (which may be stochastic) which the system follows to go from one state to the next. The collections of the states, their transition rules, and the transition probabilities form the epsilon machine itself. Once the machine is reconstructed, it can be used in a predictive fashion to predict/reconstruct the data series. Shalizi and Crutchfield [21] also prove that the epsilon machine provides the most compact representation of the dynamics generating a time series, with similar ramifications for data compression and storage as arose from consideration of the iterated function systems.

What we wish to do is consider the application of this concept to spatial patterning, particularly with reference to geological systems.

The degree of accuracy to which data can be predicted/reconstructed by an epsilon machine is inherently related to the entropy of the data. This is what we focussed on in this first excursion into applying the concepts of computation mechanics to geological data obtained by SiroVision (Figure 6).

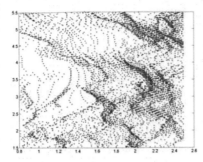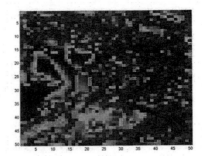

Figure 6: (left) SIROVISION data. (right) A rectangular subset of a larger data set has been selected for our experiment. Azimuth data from all sampling locations have been interpolated on a regular grid. Lack of data on the left hand side forced us to interpolate. The interpolated area is clearly visible in the smooth blue region on the left.

The hypothesis goes that if we have an infinite amount of data characterising our geological process (in this case a three dimensional image representing the SiroVision (x, y, z) data), then we can calculate the 'entropy density' (E_d), which represents the amount of unpredictability inherent in the system [6]. Put slightly differently, E_d is the uncertainty for a pixel value (averaged over all the image's pixels), given all the information about the rest of the image (Figure 7).

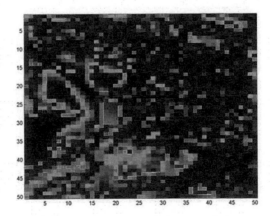

Figure 7: E_d is the average uncertainty on the pixel value, given all the information about the rest of the image.

In the real world applications, we have only finite information available. In this case, we can hope to obtain an estimate of E_d by calculating the limit entropy E for increasing amounts of data, M, and calculating/approximating the limit of E_d for $M \to \infty$, that is

$$E_d = \lim_{M \to \infty} E(M).$$

Figure 8 illustrates the process followed in determining E_d. Obviously, E(M) is monotonically decreasing. Consequently, for each M, $E(M) - E_d \geq 0$. This difference represents the amount of uncertainty in the data due only to our lack of available information. In other words, this is the amount of uncertainty we could remove if we had more data. This is an important concept in itself, because it can help us to quantify the amount of *information* we could get by collecting more *data*, and whether such collection is worthwhile at all. In the latter sentence, notice the difference in the usage of *data* and *information*.

Feldman & Crutchfield [8] define as *excess entropy* the quantity

$$E_e = \sum_{M=1,2\ldots\infty} E(m) - E_d.$$

They claim this quantity is a better measure of data structure than other techniques, mostly based on Fourier methods. To see why, we notice that E_e is not a parameterised function, that is, it does not depend on any parameter set *a priori*. This means that whatever structure E_d can detect, is does not depend on predetermined patterns or frequencies.

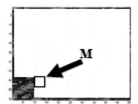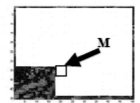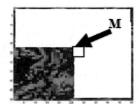

Figure 8: Calculation of $E_d = \lim_{M \to \infty} E(M)$ for increasing amounts of data M. This gives the average uncertainty on a pixel value, given an increasing amount of information on the image.

Two results came out of our experiment. First, in our SiroVision data E_d tends to zero. That is, given the information about the entire map at the resolution we decided to work on, no average uncertainty remains in the definition of the azimuth at a pixel after all data is processed. Second, we divided our image into 2 regions (see Figure 9) and calculated E_e (excess entropy) for each area. These are respectively 1.8 (bytes per pixel) for the left hand side and 1.65 for the right hand size. The left hand size seems to be less 'predictable' then the right which is most likely due to the presence of the interpolated data which do not follow the same structure as the rest of the image.

Figure 9: The SiroVision image divided into two regions. The left hand region has a higher excess entropy, most likely due to the presence of the interpolated area.

We can use E_e, the excess entropy, to determine the amount of uncertainty, structure, complexity, or randomness in an image.

4 Issues

Before we can confidently apply the latter two approaches to geological data a number of issues need to be addressed:

- Epsilon machine techniques and delayed coordinate embedding were developed to deal with 1D time series. At the next interesting level of complexity, we aim to apply these techniques to 2D spatial data. The main difficulty is that in 1D time series causality is clearly defined (event at time *n+1* follows event at time *n, i.e.,* event at time *n* can not be caused by event at time *n+1*). Spatial data may be the result of multiple processes with different directionality and accordingly causality may be hard to unravel. Interestingly, this is a conceptual problem which needs to be addressed to go from 1D to 2D. Once this is "solved", going to nD (n>2) would involve mostly computational, not conceptual, problems.

- The data are required to be on a regularly spaced grid.

- The data are required to be exhaustive; which as we have seen is difficult to attain with geological data (Figure 10).

- The data required for this analysis are discrete; the analysis of continuous data at discrete times is not yet implemented; continuous-valued, continuous-time processes are a real problem [22].

These are some of the issues our future research will address.

Figure 10: Three-dimensional joint structure in Heavitree Quartzite in Central Australia, together with eucalyptus trees, spinifex, and shadows.

5 Conclusions

-   Methods for obtaining useful geological data sets continue to improve, and it is now possib le to collect data in three dimensions in a routine manner.
-   'Computational mechanics' as an approach is very promising, but is not yet sufficiently advanced to deal with geological data sets, which are typically 3D, irregularly spaced, and incomplete.
-   Our aim of finding the causal, dynamical structure intrinsic to the process we are investigating from the patterns the process gives rise to is not yet proven impossible.

Acknowledgements

We thank George Poropat and Steve Fraser for the data set, and Paul Maconochie for his encouragement in pursuing the geological issues driving the development of SiroVision.

Bibliography

[1]   Bardsley, W.E., Major, T.J. and Selby, M.J. Note on a Weibull property for joint spacing analysis. *Int. J. Rock Mech. Min. Sci. & Geomech. Abstr.,* Vol. 27, 133-134, 1990.

[2]   Barnsley, M.F. Fractals Everywhere, p. 396. Academic Press, New York, 1988.

[3] Barton, C.C. and Larsen, E. Fractal geometry of two-dimensional fracture networks at Yucca Mountain, Southwestern Nevada. In Proc. Int. Symp. Fundamentals of Rock Joints, Bjorkliden, Sweden (Edited by O. Stephansson), pp. 77-84, Centek, Lulea, Sweden, 1985.

[4] Barton, C.A. and Zoback, M.D. Self-similar distribution of macroscopic fractures at depth in crystalline rock in the Cajon Pass scientific drill hole. In Rock Joints (Edited by N. Barton and O. Stephansson), pp. 163-170. Balkema, Rotterdam, 1990.

[5] Cheung, C.C. and Ord, A. Use of stereo shape measurement in industrial applications: passive or active? In: R.A. Jarvis, editor, Proceedings Workshop on Computer Vision - from Cognitive Science to Industrial Automation, Sydney, Australia. IJCAI-91, 12th International Joint Conference on Artificial Intelligence. Aug. 1991.

[6] Crutchfield JP, Farmer JD, Packard NH, Shaw RS. Chaos. *Scientific American*, 255: no. 6, 38-49, 1986.

[7] Crutchfield, J.P. The calculi of emergence: Computation, dynamics, and induction. *Physica D* Vol. 75, pp. 11-54, 1994.

[8] Feldman, D.P., Crutchfield, J.P.: Structural information in two-dimensional patterns: entropy convergence and excess entropy. Santa Fe Institute Working Paper 02-11-065, 2002.

[9] Hobbs, B.E. The significance of structural geology in rock mechanics. Chapter 2 in Comprehensive Rock Engineering. Vol 1, 25-62. Editors, E. Hoek, J. Hudson, E.T. Brown. Pergamon Press, 1993.

[10] Hobbs, B.E., Mühlhaus, H.B. and Ord, A. The fractal geometry of deformed rocks. Mitt. aus den Geol. Inst. ETH Zürich, Neue Folge, 239b, 28, 1991.

[11] Hobbs, B.E., Means, W.D. and Williams, P.F. An Outline of Structural Geology. 571 pp. John Wiley and Sons, Inc., New York, 1976.

[12] Hoek, E. and Bray, J.W. Rock Slope Engineering, 2nd edition, IMM, London, 1977.

[13] Hudson, J.A. and Priest, S.D. Discontinuities and rock mass geometry. *Int. J. Rock Mech. Min. Sci. & Geomech. Abstr.*, Vol. 16, pp. 339-362, 1979.

[14] ITASCA, FLAC Fast Lagrangian Analysis of Continua User's Guide, Minneapolis, Minnesota, USA, 2nd edition, 2000.

[15] La Pointe, P.R. and Hudson, J.A. Characterisation and interpretation of rock mass joint patterns. *Spec. Pap. Geol. Soc. Am.*, Vol. 199, 1985.

[16] Ord, A.: The fractal geometry of patterned structures in numerical models for rock deformation. In: J.H. Kruhl, editor. Fractals and Dynamic Systems in Geoscience, Springer-Verlag, Berlin. pp. 131-155, 1994.

[17] Ord, A. and Cheung, L.C. Image analysis techniques for determining the fractal dimensions of rock joint and fragment size distributions. In: G. Beer, J.R. Booker and J.P. Carter, editors. Computer Methods and Advances in Geomechanics. Proc. 7th Intl. Conf. on Computer Methods and Advances in Geomechanics, Balkema, Rotterdam. pp. 87-91, 1991.

[18] Packard NH, Crutchfield JP, Farmer JD, Shaw RS. Geometry from a time series. *Phys. Rev. Lett.,* 45: 712-716, 1980.

[19] Roberts, G. and Poropat, G. Highwall joint mapping in 3-D at the Moura Mine using SIROJOINT. In Bowen Basin Symposium 2000 Proceedings, Rockhampton, 22-24 October, 2000, Ed. J.W. Beeston, pp. 371-377.

[20] Rouleau, A. and Gale, J.E. Statistical characterisation of the fracture system in the Stripa Granite, Sweden. *Int. J. Rock Mech. Min. Sci. & Geomech. Abstr.,* Vol. 22, pp. 353-367, 1985.

[21] Shalizi, C.R., Crutchfield, J.P.: Computational mechanics: pattern and prediction, structure and simplicity, *Journal of Statistical Physics,* Vol. 104, pp. 817-879, 2001.

[22] Shalizi, C.R., Shalizi, K.L., Crutchfield, J.P.: An algorithm for pattern discovery in time series, *Journal of Machine Learning Research,* 2003.

[23] SiroVision. a) http://www.surpac.com/products/sirovision/ b) http://www.em.csiro.au/em/newsandinformation/earthmatters/documents/earthmatter s04-08.pdf c) http://www.ceanet.com.au/pdfs/CSIRO.pdf

[24] Takens F (1981). Detecting strange attractors in turbulence. In: Dynamical Systems and Turbulence (Eds. D.A. Rand and L.-S. Young). Lecture Notes in Mathematics, 898: 366-381. Springer-Verlag, New York.

Fractal Dimension Uncertainty from Measurements: A Probabilistic Procedure

Ian Lerche

Institut für Geophysik und Geologie, Universität Leipzig, Talstraße 35, D 04103 Leipzig, Germany

e-mail: lercheian@yahoo.com

Abstract

This paper shows how probabilistic procedures can be used to determine both the mean value as well as the uncertainty on fractal dimensions from data directly. Uncertainties are couched in terms of the cumulative probability of the true fractal dimension being less than or greater than a fixed cumulative probability value. A simple application to shear zones shows how the procedure operates. In addition, uncertainties due to the box counting method itself, and the box sizes used as counters, are examined. The resolution, precision, uncertainty, and sensitivity of results can thus be addressed simply and quantitatively. The point is also made that alternative models should be evaluated for their acceptability for it can happen that a power law fractal model is not the model that best fits the available data. The amount and quality of the available data are also shown to provide uncertainties in the fractal dimension determination. All of these points should be routinely examined when one is concerned with whether or not fractal dimension determination is an appropriate vehicle to categorize interpretations of the data field for any given problem.

1 Introduction

The practical determination of the effective dimensional scaling (the "fractal" dimension) for a variety of geological processes is of major interest in attempts to see to what extent the scaling is being modified, or even determined, by the very procedures used to quantify the scaling. This paper discusses the uncertainties brought to the fractal dimension determination due to finite data sampling, finite box counting procedures, and also provides a probabilistic procedure to assess the likelihood of a scaling dimension being appropriate. The paper also considers the "inner" and "outer" measurement yardsticks and their influence on the stability and accuracy of the scaling dimension determination. An application from residual shear stress observations is used to illustrate how one uses the procedures.

2 Cumulative Probability Considerations

2.1 General Description

Suppose one has used data to determine estimates of the mean value, E, and the standard deviation, s, of a parameter of interest. Then construct

$$\mu^2 = ln\ (1+s^2/E^2) \tag{1}$$

There are three cumulative probability points that are then easily calculated. These values are *P(16)*, *P(50)* and *P(84)*, with *P(50)* occurring at $x_{50} = E exp(-\mu^2/2)$, *P(16)* occurring at $x_{16} = x_{50} exp(-\mu)$, and *P(84)* occurring at $x_{84} = x_{50}\ exp\ (\mu)$. Here *P(16)*, *P(50)* and *P(84)* are the cumulative probability percentages of obtaining a value of *x* less than x_{16}, x_{50}, and x_{84}, respectively.

On log probability paper a plot of *x* on the ordinate versus cumulative probability on the abscissa is a straight line as shown in Figure 1.

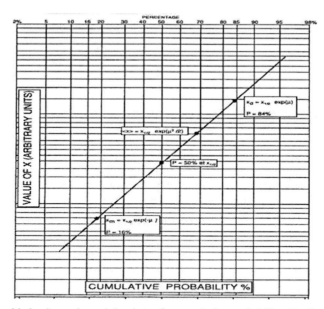

Figure 1: General behavior and special points of a cumulative probability distribution that is approximately log normal.

Because all scientific data have errors and uncertainties, and because all parameters inferred from data therefore also contain in some way the same error information, the scaling fractal dimension, *D*, must also have uncertainties. Hence one can provide a cumulative probability that the fractal dimension lies less than or greater than any particular chosen value. This point will be brought out in greater detail in the application section of this paper.

2.2 Scaling Dimension Considerations

In a shear zone, the number of faults, $N(r)$, in a square of side r is taken to scale as

$$N(r) = N_o \, r^D \tag{2}$$

where D is the scaling (fractal) dimension and N_o is a constant representing the density of fractures.

Suppose then that one has made measurements at different square sizes of the number of squares needed to cover the data field of fractures, thereby providing the suite of measurement pairs $\{N_i, r_i\}$ in $i = 1,...,$ M. Because the constant N_o enters equation (2) in a linear manner, it follows that, for any value of D, the least squares solution for N_o is given by

$$N_0 = \sum_{i=1}^{M} N_i r_i^{D} \Big/ \sum_{i=1}^{M} r_i^{2D} \tag{3}$$

Then construct from equation (2) the non-linear least squares control measure

$$X(D)^2 = M^{-1} \sum_{i=1}^{M} \{D - \ln[N(r_i)/N_0(D)]/\ln(r_i)\}^2 \tag{4}$$

with $N_o(D)$ given by equation (3). Systematic variation of D in equation (4) then produces a curve of $X(D)^2$ versus D that has a minimum occurring at say D_o, as shown schematically in figure 2. Then D_o represents the value of the fractal dimension that most closely satisfies all the data field. In addition, the value $X(D_0)^2$ then represents the average variance around this best fit. By identifying D_o as the estimate of the mean value and $X(D_0)^2$ as the variance one then has the practical values

$$D_{50} = D_0 \Big/ \big[1 + X(D_0)^2 / D_0^2\big]^{1/2} = D_0 \exp(-\mu^2/2) \tag{5a}$$

$$D_{16} = D_{50} \exp(-\mu) \tag{5b}$$

$$D_{84} = D_{50} \exp(\mu) \tag{5c}$$

so that one has a measurement-based procedure for determining the range and uncertainty of a fractal dimension directly and solely from measurements.

96

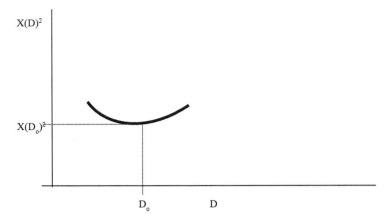

Figure 2: Sketch of the variation of the non-linear least squares measure of the text used to determine the best-fit fractal dimension and its uncertainly.

3 An Illustrative Application

The general procedure given in Section 2 has been applied to a comparison of residual structures in shear zones of different magnitudes as shown in figure 3.

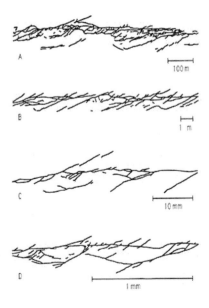

Figure 3: Comparison of residual structures in shear zones (after Tchalenko, 1970).

The four parts of figure 3 are presented on the same picture scale, although the measurement scale bars range from 100 m, 1m, 10 mm, and 1mm, respectively, as shown. Figures 3B-D refer to plasticene models of sheared zones while figure 3A is a sketch of real shear zones (Tchalenko, 1970).

For each part of figure 3 a box-counting procedure was used, with the corresponding $N(r)$ versus r results shown in figures 4A-4D. Superposed on figure 4 are the corresponding lines for best-fit values of the fractal dimension (in the sense of Section 2). Note the differences between the real rock result ($D_0 = 1.53$) and the plasticene synthetic rocks ($D_o = 1.34, 1.33$ and 1.38 respectively) in terms of the estimated best-fit fractal values over the scale range of 1-10 for r.

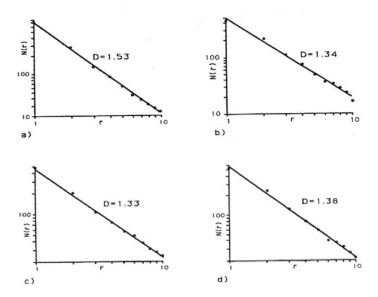

Figure 4: (a) Capacity dimensions obtained from the various shear zones: A corresponds to figure 3A, B to figure 3B, C to figure 3C and D to figure 3D.

In addition, the procedure of Section 2 allows one to determine the uncertainty on the fractal dimension estimate. One obtains

$$D = 1.53 + \begin{pmatrix} 0.01 \\ -0.03 \end{pmatrix} \text{ for figure 3A;}$$

$$D = 1.34 + \begin{pmatrix} 0.03 \\ -0.02 \end{pmatrix} \text{ for figure 3B;}$$

$$D = 1.33 + \begin{pmatrix} 0.01 \\ -0.01 \end{pmatrix} \text{ for figure 3C;}$$

$$D = 1.38 + \begin{pmatrix} 0.02 \\ -0.04 \end{pmatrix} \text{ for figure 3D;}$$

where, for instance, the value 1.53+0.01 means there is an 84% chance the actual value of D will be less than 1.54, while the value 1.53-0.03 means there is only a 16% chance the actual value of D will be less than 1.50. In this way one can categorize the uncertainty on the fractal dimension directly from the measurement field.

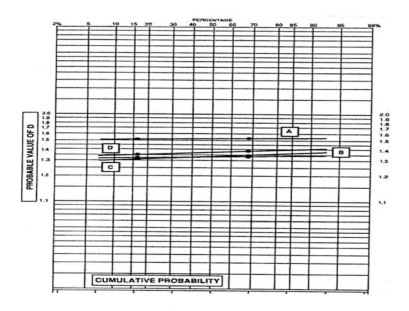

Figure 4: (b) cumulative probability plots of the various ranges of capacity dimensions for each of the four cases.

4 Box Counting Uncertainties

The size of the boxes used to obtain the estimates of mean fractal dimension and its probable uncertainty influence directly the estimates made. Figure 5A shows the estimated fractal dimension (the so-called capacity dimension) and its 16% and 84% ranges as the box size is changed for the real rock shear zones of figure 3A. The black band with white circles and error bars shows the effect of decreasing the number of points from 1-10 mm to 1-3 mm, indicating that the determination of the mean dimension and its uncertainty are relatively uninfluenced and so stable numerically. However, if the number of points is decreased in an "upward" scaling sense from 1-10 mm to 8-10 mm, then not only is there considerable uncertainty on the determined fractal dimension (as shown by the dashed lines on figure 5) but also the estimated mean value itself varies from almost 1.6 to 1.2, representing considerable instability of the determination as a consequence of the box size being too large to provide sufficient statistics to allow accurate determination of the fractal dimension.

The same sort of effect is also seen for the synthetic plasticene model of figure 3B, as shown in figure 5B, but now with even larger fluctuations of the mean estimate and also of the 16% and 84% cumulative probability values because, by direct inspection of figures 3A and 3B, there is a lower total count of fractures for the synthetic material. This point has been eloquently addressed by Essex (1991), who has noted that it allows the sensitivity of the system to be tested, especially when the number of data is small and, at a larger box scale, the amount of data must eventually become insufficient to provide an adequate representation.

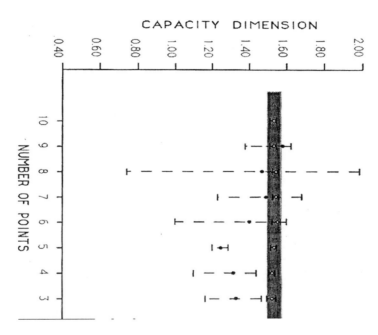

Figure 5: (a) Data from the shear zones of figure 3A; Best estimates of D and its range of uncertainty versus the number of points taken into account in the determination.

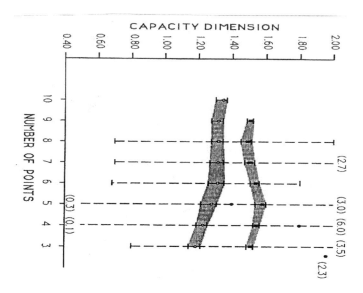

Figure 5: (b) Data from the synthetic plasticene model of figure 3B. Solid lines: range of measurements decreasing from 1-10 mm to 1-3 mm; dashed lines: range of measurements decreasing "upwards" from 1-10 mm to 8-10 mm.

5 Discussion and Conclusion

The cumulative probability procedure given allows one to quickly and accurately assess the resolution, precision, and sensitivity of a fractal dimension directly from measurements. While the procedure has been illustrated only for a simple example of shear zones, it is this very simplicity that allows one to clearly and cleanly focus on the basic problems. More complex situations can be handled in similar fashion.

In addition, it is <u>not</u> clear that the model choice of a power law dependence for scaling is the best model that could be chosen. For instance, a Weibull distribution choice of the form

$$N(r) = N_0 \left\{ 1 - \exp\left[-\left(r/r_0 \right)^D \right] \right\}$$

which reduces to $N(r) = N_0(r/r_0)^D$ for $r \ll r_0$, but which is very different than the power law behavior at $r \gg r_0$, is an alternative possibility, as are many other models. For each model chosen one must devise linear and non-linear procedures, equivalent to that given in Section 2, to test the model behavior against observations, including uniqueness, precision, resolution and sensitivity.

Further, the stability of any fractal dimension determined by such procedures is dependent on the data quality and quantity, on the box sizes used for the determination, and on the number of counts ensuing at different box sizes. The interpretation of a fractal dimension must carry with it both the model assumptions made, as well as the uncertainties arising from the methods and procedures used to determine the dimension.

Only in this way can one be confident of the worth of such interpretations. It is these points that the quantitative procedure and the application have been tailored to illustrate as simply as we know how. It is to be hoped that similar procedures will now be more routinely applied to fractal dimension determination than seems to have been the case to date.

Acknowledgements

This work has been partially supported by the DAAD through their award of a Visiting Professorship to I.L. at the University of Leipzig, which is also thanked for its contribution to this support. Professor Werner Ehrmann is particularly thanked for the courtesies and support he and his group at Leipzig have made available during the course of this work. Olivier Bour contributed substantially to this work and he is thanked for his assistance.

Bibliography

[1]   Essex, C., 1991, Correlation dimension and data sample size. In Nonlinear Variability in Geophysics: Scaling and Fractals (Schertzer, D. and Lovejoy, S., eds.), Kluwer Academic Press, Dordrecht, p. 93-98.

[2]   Tchalenko, J. S., 1970, Similarities between Shear Zones of Different Magnitude, Geol. Soc. Am. Bull., 81, 1625-1640.

Characterization of anisotropy – scale relations for complex irregular structures

Department of Geography, Saint Mary's University, 923 Robie St., Halifax, NS, B3H 3C3, Canada

e-mail: cristian.suteanu@smu.ca

Abstract

The paper presents a methodology for the study of anisotropic aspects in the case of complex structures. The proposed multiscale approach relies on percolation properties of a "virtual fluid" that is applied to the studied structures, after superposing them with a channel grid with variable orientation. The described methodology is designed to address natural structures that include numerous elements, with complex configurations (curved, interrupted, with variable width, intersecting other elements etc.), in order to characterize anisotropy and its relation to spatial scale. Areas of applicability include fault sets and fractured surfaces, fragmented rock systems at different scales and their fluid flow properties (with implications for groundwater and groundwater pollution studies), but they may also concern very different features like sand dunes and their spatial dynamics.

1 Aim of the paper

Complex natural structures studied in the framework of geomorphology, geography, geology, geodynamics, as well as of geotechnical engineering, are usually characterized by a large spatial variability. The quantitative characterization of their irregular geometry, albeit very important for many applications, often represents a challenging undertaking.

Anisotropy is a parameter that proves to be among the most difficult to describe. The number and the variety of geometric elements involved in numerous situations, as well as the complex way these elements may be interconnected, make it difficult to characterize even common images of fractured rock structures corresponding to a certain scale. However, the fact that anisotropy may change with the considered spatial scale makes this task even more problematical. Nevertheless, the significance of anisotropy and its practical implications require methods capable to deal effectively, in a relevant way, with the many interacting aspects contributing to – or affected by – anisotropy.

The present paper addresses the challenge of characterizing anisotropy, as well as its relation to spatial scale, in the case of complex irregular structures.

103

2 Challenges and principles

Under these circumstances, our endeavor is both supported and stimulated by the effervescent climate created by a fast developing nonlinear science [1]. Beyond a rich methodology, the latter provides a series of important principles, which have emerged from and are confirmed in a multitude of disciplines. These principles concern features that have more and more in common with the "real world": the systems are complex, made of parts that interact in a way that is nonlinear, their dynamics is often sensitively dependent on initial conditions, their properties emphasize special relations to the observation scale. "Analyzing" them by separating them from the surrounding "context" is not always helpful: long-range correlations may be at work, and an investigation that does not consider the interactions between the system and its environment at different scales may fail to understand the underlying dynamics. The search for the "best" scale to study the system may not always be a helpful idea either. Considering the system at different scales, and trying subsequently to extract correlations from the different views corresponding to the whole range of scales, proves to be a more effective approach [2]. This is particularly relevant when dealing with complex, irregular structures, while many natural systems – if not most of them – belong to this category. In our attempt to produce a methodology for the rigorous characterization of anisotropic aspects, grasping also their relation to spatial scale, we shall thus rely on a multiscale approach, considering also connections and interactions between the different parts involved in the studied processes. Multiscale methods prove to be very effective when applied to complex geological structures [3]; dynamic aspects like the earthquake activity in a seismogenic area can also be analyzed effectively using a multiscale perspective [4].

We shall start from the premise that – in today's context – getting information about the systems we are interested in is not enough. In fact, due to the spectacular advances in technology, information is often produced generously. However, this does not mean that the challenges posed by complex systems are solved. From the information to be processed, meanings must be reached: paradoxically, the task of detecting meanings is not simplified when the information quantity is increasing (as one would be tempted to expect): on the contrary, the fact of dealing with huge quantities of information, far from facilitating our understanding, has become itself a real challenge. What we need is not just information: it is *structured* information. Connections between different pieces of information, correlations in the data, represent powerful ingredients for information structuring. On the other hand, relevant information is not very useful, unless it is represented properly. Sometimes, the complexity of the problem addressed may lead to graphical forms (or multiple graphical layers) that are hard to interpret. The examination of the output produced by a certain method (diagrams, graphs) may be a challenge in its turn: the design of a new methodology should thus consider also this kind of problems; it should provide outputs able to offer insights with respect to relations between parameters, and it should present the information in a readily interpretable manner.

In other words, it is desirable to detect rigorously relevant correlations, and to communicate them in a useful way.

The envisaged areas of applicability are relatively wide. They include fault sets and fractured surfaces, fragmented rock systems at different scales and their fluid flow properties (with implications for groundwater and groundwater pollution studies), but they may also concern very different features like sand dunes and their spatial dynamics.

3 The virtual fluid flow approach

When addressing the problem of fluid flow through fractured rock structures, it is important to note that transport properties of rocks depend crucially on the rock pore structure and on the geometry of fractures; a special role is played by connectivity properties. We usually deal with a particularly strong spatial variability, with distinct features on various scale ranges. Anisotropy obviously plays a key role for the behaviour of rock systems, and it is essential to have tools to evaluate it quantitatively. Our goal will thus be to characterize the complex geometry of fractured rock structures, with a special concern for scale and orientation dependent properties.The flow of a fluid through fissured media is extremely sensitive with regard to anisotropy and connectivity properties of the fissures [5]. Therefore, by observing the behaviour of a fluid in a set of fractures, one would acquire precious information about these important aspects of a rock system.

Relying on this idea, we can use the advantages of a fluid flow experiment, by applying – not a real, but a virtual fluid, and not necessarily to a fractured rock system, but also to other different structures, for which we want to investigate anisotropy and connectivity aspects [6].

Despite the more general applicability of the proposed methodology, in the following sections we shall use, for convenience, a language that refers specifically to fractured rock systems. We shall thus mention "fissures" and "fractures", which does not mean that the effectiveness of the described procedures is limited to the domain of fractured rock. Nor does this imply that only a certain range of scales can be addressed, since – as it will become evident – the method is meant to provide useful results, for instance, also at the scale of pores.

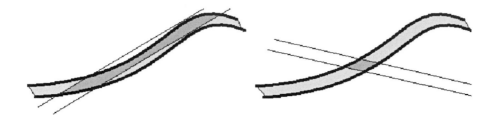

Figure 1: Virtual fluid flow confined by a channel with different orientations.

4 Percolation lengths: anisotropy vs. scale

To evaluate the anisotropy of a set of fissures with the help of virtual fluid flow, we start by superposing a "channel" of a given width to the image of the fissured surface. The virtual fluid is then allowed to "flow" in the fissures, confined by the limits of the applied channel. The effect of the flow can thus reveal anisotropic aspects of the fissure structure.

For instance, even if we apply this procedure to a fissure that is different from a straight line, the length over which the fluid can flow is strongly dependent on the direction of the applied channel (Fig. 1). The "percolation length" will thus reveal preferential spatial orientations of the studied structure.

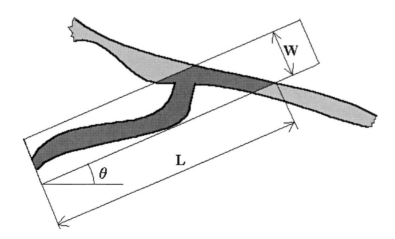

Figure 2: Virtual fluid flow in the case of a curved, bifurcating structure.

This procedure can be successfully applied also in the case of complex sets of fractures, including fractures with variable width, bifurcations, intersecting other fractures etc. For each orientation angle θ and for each channel width W, we can measure a percolation length L (Fig. 2).

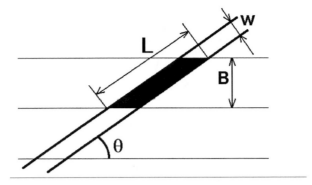

Figure 3a: The virtual fluid flow procedure applied to a grid of parallel, equally-spaced "fractures": main elements.

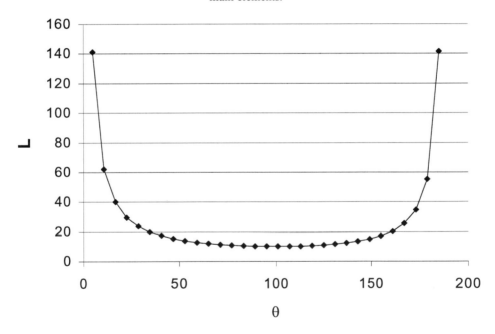

Figure 3b: Variation of length L (arbitrary units) with channel orientation (fracture width B=10 units).

To see the effects of the channel orientation for the measured percolation lengths, we may consider the simple case of parallel, equally spaced "fissures" (Fig. 3a). For a fissure width equal to B, a channel of width W, smaller than B, will lead to a percolation length that depends on the orientation angle θ:

$$L = \frac{W \cos\theta + B}{\sin\theta} \qquad (1)$$

From here we can see that the length L increases quickly when the channel orientation angle θ gets close to the orientation of the fracture direction (in this case, 0° or 180°, see also Fig. 3b). The fast growth of the percolation length for certain angle domains signal the existence of a long fissure corresponding to this orientation.

To avoid difficulties involved by situations in which the length tends to grow to an infinite value, a limit L_m for the maximum length is set: thereby, every time the percolation length becomes larger than this limit, its value is conventionally set to L_m.

The scale dependence of the structure, from the point of view of anisotropy detected by fluid percolation, can be revealed by an evaluation that uses not one, but a series of values of the channel width W. A scaling relationship may then be found, as long as $L < L_m$:

$$L \sim W^{\xi} \qquad (2)$$

where the exponent ξ characterizes the structure for the given orientation θ. In other words, a value $\xi(\theta)$ may be identified for each orientation θ (with a step $\Delta\theta$).

In practice, however, such a relationship may be less useful: the interval over which this relationship can be established is significantly limited $(L < L_m)$.

On the other hand, even if eq. (2) does not hold, the relation between scale and percolation length is still interesting. In fact, for real fissure patterns, the $L(W)$ curve may get saturated before L reaches the limit L_m. Figure 4 illustrates this situation for a crack pattern (circles in the diagram), compared to a set of parallel, equally spaced fissures, when the channel orientation coincides with the direction of the parallel fissures (squares in the diagram). In the latter case, no matter how small the value of the width W is, the percolation length tends to be infinite, and it is therefore set to L_m. The saturation phenomenon seen in the case of the crack pattern (circles in Fig. 4) is relevant for our understanding of the studied pattern. Therefore, instead of leaving it out (as we would do by focusing on the scaling process alone), we will want the analysis to provide the information concerning the saturation: the most interesting aspect is represented by the *scale* at which saturation sets in.

To this end, we represent – for each orientation – the successive slopes that characterize the intervals in graphs like those in Fig. 4.

Due to the spatial variability of natural crack patterns, it is not enough to study the L-W relation for a single position of the percolation channel. The latter should be located in all possible positions on the studied fracture network.

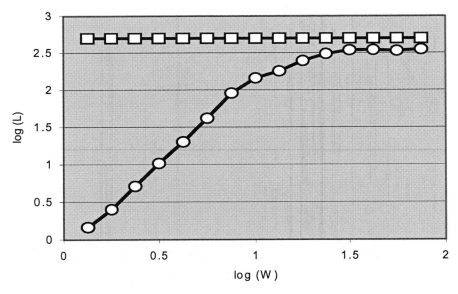

Figure 4: Percolation length vs. channel width in the case of a crack structure (circles) and for a set of parallel, equally-spaced fractures when the channel orientation is the same with the orientation of the parallel fractures (squares). The level of the squares corresponds to $\log(L_m)$.

The percolation length for the channel width W is then found as the average value for all tested positions. In practice, the number of positions can be limited: if we denote one of the axes by X and the other by Y, then the minimum increments along these axes can be equal to 1/3 W.

We characterize thus the pattern by a "local scale range parameter" D_L, which reflects structure properties that do not belong to a scaling pattern over the whole studied scale interval – they are "local" from the point of view of the scale range (a concept introduced in [6], where it was defined in a different way). So, D_L is:

$$D_L(i,\alpha) = \left\langle \frac{\log L_{i+1}(\alpha) - Log L_i(\alpha)}{\log W_{i+1} - \log W_i} \right\rangle_{X,Y}$$

(3) where the symbol $< .. >$ shows that the values are averaged over all $X,\ Y$ positions.

To represent this information according to the principles stated in section 2, we should include in our diagrams the orientation angle, the considered spatial scale range for each point on the L-W graph, as well as the corresponding local scale range parameter.

A possible representation consists in choosing for the diagram coordinates the scale range and the orientation angle, and to plot the local scale range parameter according to a colour code. Based on the local scale range values, we can thus produce contour maps, which lend themselves to an effective interpretation. Figure 5

5 shows an example of such a contour map for a synthetic fault set model. The colour code used in the diagram is specified in its upper bar. The diagram reflects a variety of situations that depend on the orientation angle and on the scale interval, from smooth to tortuous percolation paths. As a consequence, it is possible to assess a complex structure from the point of view of its anisotropy, as related to scale.

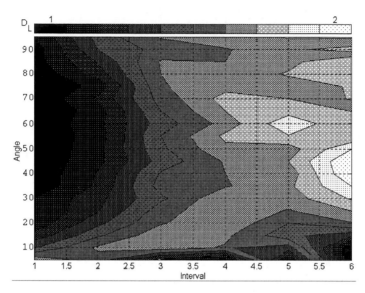

Figure 5: Local scale range contour map. Abscissa: scale range in logarithmic coordinates. Ordinate: orientation angle. The colour code (hatch pattern) reflects the tortuosity of the percolation path.

Alternatively, instead of the individual scale intervals, one may consider scale ranges of various lengths, characterized by a stronger or a weaker linear correlation of the points in the L-W graph, and to draw isocorrelation maps [7].

The strong spatial variability of natural structures like those mentioned in this paper suggests that the multifractal formalism should be effective. We therefore consider the application of a robust method for the determination of the multifractal spectrum. Such a robust method was introduced by Chhabra and Jensen [8]: however, despite its advantages with respect to reliability and accuracy, the method is not orientation sensitive. Based on the described virtual fluid flow procedures, we may change the method and make it relevant for anisotropy characterization [6]. A brief description of the method follows below.

In a first step, we cover the structure to be analyzed with a grid of channels of width W, having an orientation angle θ. Virtual fluid is then injected in all the fissures along the channels of the grid. The percolation lengths will vary widely with the characteristics of the fracture pattern. We can use now these lengths in order to codify the segments of the structure: all the segments will be coloured with a codified pattern, corresponding to their percolation features (Fig. 6).

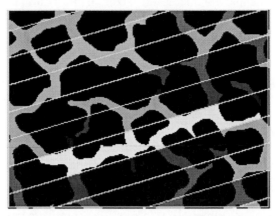

Figure 6: Percolation channel grid applied to the analyzed structure. Fracture segments are codified (and coloured correspondingly) depending on their percolation length.

For every channel, we compute now the probabilities P_i corresponding to these weighted percolation segments:

$$P_i = A_i c_i \qquad (4)$$

where A_i are the segment areas and c_i are their corresponding percolation codes. The probabilities P_i are then normalized to the unity. Starting from these probabilities, we can compute now the moments of order q according to the method of Chhabra and Jensen [8]:

$$\mu(q,W) = \frac{[P_i(W)]^q}{\sum_{j=1}^{N}[P_j(W)]^q} \qquad (5)$$

where N is the number of channels corresponding to the channel width W and the orientation angle θ. We thereby amplify, on one hand, the denser regions (when $q > 1$), and on the other hand, also the less dense regions (when $q < 1$). With these data, we compute the following implicit functions of q:

$$f(q) = \lim_{W \to 0} \frac{\sum_{i=1}^{N} \mu_i(q,W) \log \mu_i(q,W)}{\log W} \qquad (6)$$

$$\alpha(q) = \lim_{W \to 0} \frac{\sum_{i=1}^{N} \mu_i(q,W) \log P_i(q,W)}{\log W} \qquad (7)$$

The procedure is repeated for increasing sizes of the channel width W. For each value of q, one can obtain thus a value for f and one for α, from the linear relations $\Sigma\,\mu_i \log \mu_i$ vs. $\log W$, and $\Sigma\,\mu_i \log P_i$ vs. $\log W$, respectively, which leads to the $f(\alpha)$ spectrum.

The same procedure is then applied for each value of the orientation angle θ (again, with a step $\Delta\theta$). An $f(\alpha)$ curve is thus produced for each orientation step. The resulting diagram (Fig. 7 presents an example) consists of a multitude of $f(\alpha)$ curves, which may reveal angle domains characterized by a narrow or a less narrow spectrum of exponents, by different values for the maximum of the curve (corresponding to the "box dimension") etc. The main problem implied by this methodology stems from the fact that the available data (especially the scale domains) may not always ensure a proper definition of the singularity spectrum. However, when the spectrum can be defined, it provides a reliable picture of the scaling aspects of the structure, related to spatial orientation.

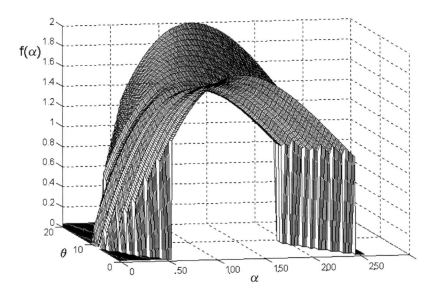

Figure 7: Example of a set of multifractal spectra $f(\alpha)$ corresponding to different orientation angles θ.

5 Conclusions

The described methodology enables us to analyze many categories of natural structures – including large sets of elements (like fractures), with complex configurations (curved, interrupted, with variable width, intersecting other elements etc.), and to characterize anisotropy, as well as the relation between the latter and the spatial scale.

Although we use throughout the article a "fractured rock" vocabulary, the areas of applicability are not confined to these applications. Other, very different, geomorphological structures, like sand dunes for instance, can also be considered. The methods can be applied on a large variety of scales: from aerial photographs of geomorphological features to images of pore structure; the domain of pore-fissure interaction can thus also be addressed successfully.

A critical analysis of the virtual fluid flow procedures shows that the final results reflect a combined effect of shapes, lengths, orientation, and relative positions of the considered elements. Therefore, it is difficult to discriminate the significance of all these sourcesfor the anisotropic properties. However, on the other hand, information about this combined effect is precious: it is otherwise very difficult or virtually impossible to deduce such an effect – and the overall anisotropic behaviour of a structure – based on information about these aspects taken separately. Often, it is this overall anisotropic behaviour that we are interested in. Applications concerning real and modeled structures will reveal the strengths of the methods, as well as the aspects requiring further development.

Acknowledgements

The author would like to thank Prof. Dorel Zugravescu, Prof. Florin Munteanu and Cristian Ioana, with whom he completed the first stages of this research.

Bibliography

[1] Suteanu, C.: Not the same geoscience: sources and impacts of the change in geoscientific communication, Joint Meeting of the Atlantic Geoscience Society and the Northeastern Section of the Geological Society of America, Halifax, March 27-29,2003.[2] Kruhl, J.H. (Ed.).: Fractals and Dynamic Systems in Geoscience, Springer, New York, 1994.

[3] Ouillon, G., Sornette, D., Castaing, C.: Organisation of joints and faults from 1 cm to 100 km scales revealed by optimized anisotropic wavelet coefficient method and multifractals analysis, Nonlinear Processes in Geophysics, Vol. 2, pp. 158-177, 1995.

[4] Suteanu, C, Zugravescu, D., Ioana, C.: Dynamic fingerprints of dissipative systems with discrete appearance: Applications in the study of seismicity, Yearbook for Complexity in the Natural, the Social and the Human Sciences, Vol. XI: J.H.Kruhl and H.-J.Krug (eds.), Non-Equilibrium Processes and Dissipative Structures in Geoscience, Berlin, Duncker & Humblot, pp. 209-228, 2001.

[5] Kolditz, O.: Flow, Contaminant and Heat Transport in Fractured Rock, Borntraeger, Stuttgart, 1997.

[6] Suteanu, C., Zugravescu, D., and Ioana, C.: Fractured rock characterization: Scaling and angle distribution of percolation lengths, Proceedings, Int. Conf. "Fractured Rock 2001", Toronto, 2001.

[7] Suteanu, C., Kruhl, J.H.: Investigation of heterogeneous scaling intervals exemplified by sutured quartz grain boundaries, Fractals, Vol. 10, No. 4, pp. 435-449, 2002.

[8] Chhabra A., Jensen R.V.: Direct Determination of the $f(\alpha)$ singularity spectrum, Physical Review Letters, Vol. 62, No. 12, pp. 1327-1330, 1989.

Fractal geometry analyses of rock fabric anisotropies and inhomogeneities

Jörn H. Kruhl[1], Francisc Andries[1], Mark Peternell[1] and Sabine Volland[2]

[1] Tectonics and Material Fabrics Section, Faculty of Civil and Geodetic Engineering, Technische Universität München, Arcisstraße 21, 80290 München, Germany,

e-mail: kruhl@tum.de, francisc.andries@tum.de, mark.peternell@tum.de

[2] Chair of Engineering Geology, Faculty of Civil and Geodetic Engineering, Technische Universität München, Arcisstraße 21, 80290, München, Germany,

e-mail: sabine.volland@mytum.de

Abstract

Fabric and its inhomogeneity and anisotropy, as one of the dominant characteristics of artificial and natural crystalline material, can be mainly studied on the basis of geometrical properties, that is the pattern. Three newly developed fractal geometry methods are exemplified that represent first steps towards a quantification of pattern inhomogeneity and anisotropy and allow the comparison between different patterns and parts of patterns. Such type of quantification facilitates the comparison between naturally, technically and experimentally produced fabrics. Pattern quantification will form a future useful fundament for the analysis of fabric-forming processes. A better knowledge about such processes is important for many fields of the geo- as well as engineering sciences.

1 Introduction

Fabric is one of the dominant characteristics of artificial and natural crystalline material. We define this term as the geometric and crystallographic properties of material domains (on all scales) — shape, size, spatial distribution and orientation, crystallographic orientation and internal crystal structure. The domains representing the fabric may range from the atomic scale up to the scale of the continental crust. The fabric is formed and changed mainly by temperature, pressure (hydrostatic pressure), differential stress, strain rate, rate of stress change, time, material properties, fluids and the fabric itself. Since most properties of the fabric are geometric ones a fabric is partly, or in special cases even totally, represented by a pattern, i.e. the purely geometrical arrangement of material domains. Even if the combination of geometrical and non-geometrical characteristics of the fabric appears promising with respect to the amount and quality of information, a purely geometrical fabric analysis, i.e. a pattern analysis, is still the first and most effective step in fabric investigation — the more so as the limits of pattern analysis are not touched by far and a large variety of methods are still waiting for being explored.

A common way to analyze the structuring of rocks is based on the representation of *particle-size frequency histograms*, together with the presentation of *cumulative curves in log-log diagrams*. This is well-known since many years, e.g. from sedimentology [24]. This type of log-log diagrams also results from the application of *fractal geometry* on data sets. Cumulative particle-size frequency curves with linear trends in log-log diagrams give a first hint that the particle pattern is power-law related as is typical for fragmentation-produced particle-size distributions [32, 25, 26, 9, 17].

The development of techniques of fractal geometry during the last two decades formed the basis for modern pattern analysis. Fractal geometry offers excellent opportunities (i) to characterize complex patterns in geomaterials quantitatively, (ii) to receive information on pattern-forming processes and their interaction, and (iii) to investigate the anisotropy and the local variation of a pattern. Furthermore, the quantification of patterns allows the comparison of different processes in different geological as well as artificial environments or of natural and experimental data. In general, methods of fractal geometry test the self-similarity of a pattern and provide a quantitative parameter: the *fractal dimension D*. Amongst many different methods, the two most widely applied ones are (i) the *perimeter* (or '*compass*', '*divider*') method and (ii) the *box-counting method* or related methods, e.g. the *Minkowski's sausage logic* [13, 9, 20]. The perimeter method is particularly suitable for the investigation of linear structures, e.g. coastlines, or outlines of rocks, irregular fractures, and sutured grain boundaries, all in 2-d. The box-counting method can be applied to the 2-d investigation of more or less isotropic patterns, e.g. fracture networks, crystal or pore-space distributions, etc. Recent investigations indicated that different methods show different sensitivities and advantages in relation to the same pattern [5, 12] and that different types of patterns require different analysis methods that partly still need to be developed. Most promising are new techniques that do not investigate geometric structures but analyze the structures of data sets, generated by fractal geometry techniques, e.g. of data points in a double-logarithmic plot, as the result of the perimeter method [30]. In addition, the validity of a variety of existing methods in relation to different types of patterns needs to be tested.

In the present paper we demonstrate three examples of newly developed pattern analysis methods. The first one is related to pattern inhomogeneity, exemplified by inhomogeneous crystal distributions in a porphyric granite.

The other two methods are related to pattern anisotropy and represent a modified Cantor-dust method applied to quartz grain boundary patterns on the micro-scale and to a fracture pattern on the centimeter scale from a brittle deformed rock.

These methods are steps towards the quantification of pattern anisotropies and inhomogeneities in rocks but may be also applied to any material and on any scale.

2 Pattern inhomogeneity

The *box-counting method* [4, 9, 21, 22, and many others] is a widely used fractal geometry method, also due to its simplicity and capability of being easily automated. A grid of square boxes of side length s is superimposed on the pattern to be analyzed. The numbers of boxes that are not empty are counted. This procedure is repeated for as large a range of s-values as possible. Finally, the number $N(s)$ of the boxes transected by the pattern are plotted versus the reciprocal size $(1/s)$ of the boxes in a double-logarithmic diagram. These data points exhibit a linear correlation if the investigated pattern is fractal. The slope m of the linear correlation represents the fractal 'box-counting' dimension D_B. In a plane, D_B can never exceed 2 [7, 21, 22]. But the method does not provide information about a possible anisotropy and inhomogeneity of the pattern. This is disadvantageous because, typically, many geological as well as artificial structures are anisotropic and/or inhomogeneous.

3 Crystal distribution inhomogeneities in igneous rocks

Recently, first steps into the analysis of crystal distribution inhomogeneity have been made. Based on kriging, a modified box-counting method provides a matrix of fractal dimensions and, consequently, a quantification of the pattern inhomogeneity [23].

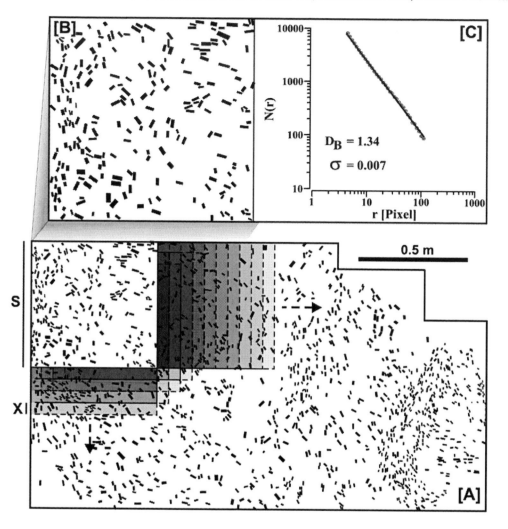

Figure 1: Map-counting method, applied to a pattern of idealized K-feldspar phenocrysts from a porphyric granite. A square of a specific side length s is shifted in two orthogonal directions over the
structure, with steps of step-length x (kriging method) [A]. The pattern within each single square [B]
is analyzed by the box-counting method. This method results in a fractal dimension D_B, determined
as the negative slope of the linear regression in a double-logarithmic plot of the box length r versus
the number of the boxes N(r) transected by any of the crystals [C]. For details see [9]; σ = standard
deviation; sketch based on photographs of a horizontal rock surface.

In porphyric granites, due to (i) inhomogeneous melt compositions, (ii) different
nucleation and crystallization rates, and (iii) flow of the melt-crystal-mush, feldspar
phenocrysts may be inhomogeneously distributed (Fig. 1 and 2). On the other hand,
the opportunity arises to analyze these variable conditions on the basis of inhomogeneous crystal distribution patterns. For example, the application of the box-
counting method on different parts of a 2-d crystal distribution pattern leads to a
specific fractal box-counting dimension D_B for each part of the investigated total

area. This requires the application of a kriging procedure. A combined photo-graphic and graphical recording of the rock section leads to a sketch of the feld-spar distribution that comprises not only the positions but also the orientation and size of the crystals (Fig. 1A). In our example, only feldspars larger than ~1 cm and with an axial ratio > 1.5 were included. This sketch forms the basis of the subse-quent application of the box-counting method. It already shows the clearly inho-mogeneous distribution and locally strong preferred orientation of the crystals.

During the kriging procedure a square of a specific size, dependent on the size of the investigated area and on the fineness of the structure, is shifted over the struc-ture in defined steps (Fig. 1A). The fractal box-counting dimension D_B is deter-mined for each position of the square (Fig. 1B,C). These D_B values, in the center of each square, form a grid of D_B numbers with the respective square side length as grid distance and a marginal cut-out of half of the square side length. The local variation of D_B can be presented on an isoline map where the isolines mark regions of higher or lower fractality of the crystal distribution (Fig. 2A). The comparison with a crystal density distribution map (Fig. 2B), constructed in the same way by a kriging procedure, shows a good match between fractality and crystal density in some regions, in others not. Such similarities are due to the fact that D_B reflects the 'density' of a structure, i.e. the degree of 'area filling'. Consequently, D_B increases with increasing crystal density. On the other hand, D_B also reflects the complexity of the structure, i.e. of the crystal distribution. This distribution is independent of the crystal density. For example, the D_B values are clearly increased in the central part of figure 2A although the crystal density is low and uniformly distributed (Fig. 2B).

As a disadvantage of this type of kriging procedure and in contrast to the density determination, the square size cannot be infinitely reduced because the box-counting method is carried out within one square and requires a certain range of different box-sizes for a statistically validated result.

However, if the box size is too small, again, the pattern within the box is too small or not complex enough for a statistically safe result, i.e. for the statistically safe determination of the fractal dimension.

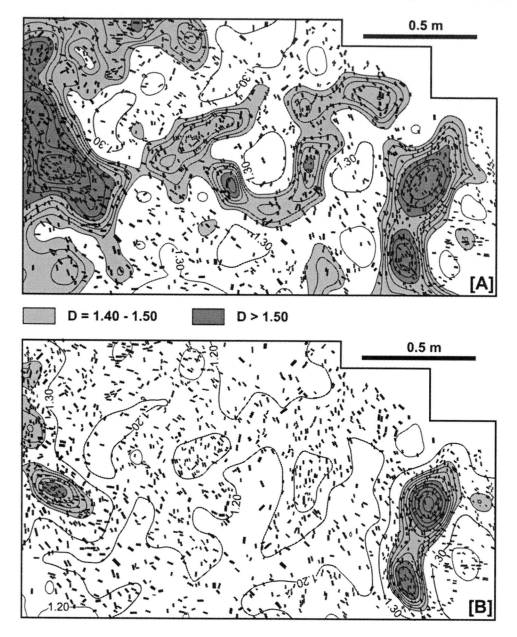

Figure 2: [A] Isoline presentation of the distribution of the fractal box-counting dimension determined by means of the map-counting method (Fig. 1). Contours above 1.4 = 0.25. [B] Isoline presentation of the crystal density distribution (number of crystals per unit area). The density values have been multiplied by the quotient of the maximum fractal dimension and the maximum density, in order to match both contouring

In addition, a large square reduces the area where isolines of the fractal dimension can be constructed. Consequently, the pattern area suitable for analysis is reduced.

4 Pattern anisotropy

The anisotropy of a pattern cannot be investigated by the 2-dimensional box-counting method but requires a technique working in 1-d. One such method of anisotropy quantification is based on the *Cantor set* [21, 22].

The classical Cantor set can be graphically constructed by the subsequent removal of sections from an initial line of length 1. The infinite repetition of the removal procedure leads to an infinite number of points, the *Cantor dust* [13]. This linear sequence of points represents a self-similar pattern. Since its volume is less than that of a line, it has a fractal dimension D, with $0 < D < 1$. Methods based on the Cantor dust offer opportunities to study the distribution of material in one dimension and were applied in geology to fractures and faults and the thickness distribution of veins [16, 11], the spacing of veins [14, 27] and fractures [2, 34, 35, 1], or to the distribution of geological events [28, 3].

Two different types of 1-dimensional analysis are generally applied, (i) the *spacing population technique* [6] and (ii) the *interval counting technique*, the 1-d equivalent of the 2-d box-counting method [34, 35]. The first one is regarded to be more suitable because (i) it receives the fractal dimension directly from a point arrangement along lines through the pattern, and not on the basis of an additional procedure, the box-counting method, as the *interval counting technique* and (ii) it can discriminate different types of planar geological structures. However, the *spacing population technique* is sensitive to truncation effects (see discussion by [5]).

In relation to a specific data set, the box-counting as well as the cantor-dust method may result in more than one fractal dimension. Such different fractal dimensions are interpreted to reflect different related or independent processes or changes of the internal structure of material, which lead to different shapes on different scales. For example, the outlines of mineral grains may lead to a particular fractal dimension and their arrangement in larger arrays may result in another fractal dimension on a larger scale, named *textural* and *structural* fractal dimension [18, 9].

5 Fracture patterns

Fractures in rocks or artificial materials are an omnipresent phenomenon. They occur from the nanometer to the kilometer scale and, therefore, represent a material fabric with an extreme variation of magnitude. Fracture patterns provide information about the conditions of their formation. (i) The orientation of parallel extension fissures gives the rough orientation of the principal extension direction.

(ii) On the basis of conjugate fractures with known sense of movement, the orientation of the principal stress axes can be determined [8]. However, more complex fracture patterns could be only analyzed after the establishment of fractal geometry. It turned out that many (but not all!) fracture patterns are self-similar or self-affine over one or several orders of magnitude and, therefore, follow a power-law. Such a relationship is generally taken as evidence for the generation of the fracture pattern by one single process and models have been developed and simulations performed to explain such patterns [15, 33, 29]. In addition, two different orders of magnitude with different types of self-similarity of the pattern are taken as indication of two different subsequent fracture forming processes [36].

Many fracture patterns are anisotropic. This anisotropy is caused by specific conditions, mainly the type of stress, and, therefore, allows interferences on these conditions. Nevertheless, anisotropy is rarely considered. As follows, we describe a fractal geometry method for the anisotropy quantification of a 2-d fracture pattern.

One section from a brecciated fracture zone in NW Sardinia have been analyzed by the *spacing population technique* on the basis of a line drawing of the rock fragment outlines [36]. Parallel lines were superposed the line drawing and oriented in 18 different directions of 10° difference each (Fig. 3). Subsequently, the lengths of the line segments in the fragments (shown in bold) are measured and their cumulative frequency distributions presented in a double-logarithmic plot, illustrated for the 60° and 140° direction (Fig. 4).

The data points can be divided in two intervals with different linear regressions and, consequently, two different slopes of the regression lines. This holds for the diagrams of all the other directions. The subdivision is based on the clear difference of the slopes $m1$ and $m2$. Despite the local scattering

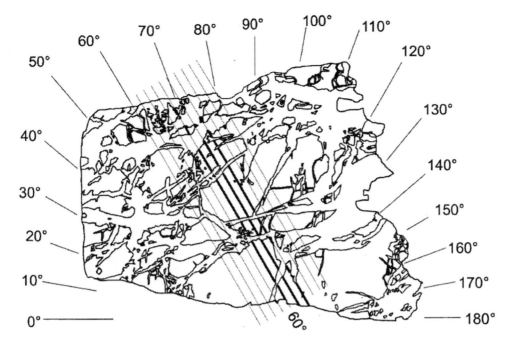

Figure 3: Measurement procedure for the construction of the line segments in the rock fragments, which form the data basis for the determination of direction-related fractal dimensions, presented in Fig. 4. Equal-spaced parallel lines with variable orientation, in this example 60° deviating from the horizontal, define segments of particular length. For three lines the segments are shown in bold. From [36].

of the data points, all correlation coefficients are near or above 0.99 and all standard deviations are (mostly far) below 0.01. Larger segments of lower frequency (below 10-20) have been excluded because their small number implies large fluctuations and therefore, does not allow an 'objective' computation in the large-size domain. The deviation of the data points, including the shortest segments (below 1.2 mm) from the regression line to lower frequencies, is most probably due to the fact that such small segments are more easily missed during the measuring procedure than larger segments.

The high correlation coefficients, the low standard deviations and the clear subdivision into two slopes support the view of a general linear regression over about 0.5 to 1 order of magnitude. Such small intervals of linear regression are typical for natural geological structures and do not argue against the validity of such correlations. The slopes of the regression lines define the fractal dimension D of the segment measurements. Such different

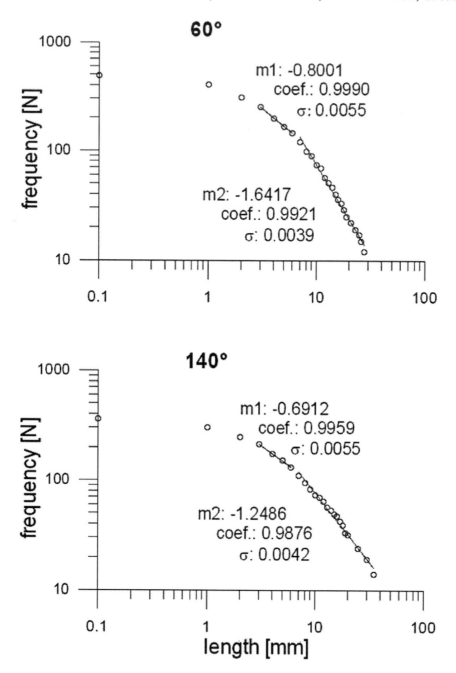

Figure 4: Double-logarithmic plot of segments formed by two sets of lines of 60° and 140° orientation, superposed the line drawing in figure 3. The data points of the two diagrams can be divided in
two groups with different linear regressions and, consequently, two different slopes (*m*1 and *m*2).
The lengths of the correlation lines span the data points included by the logarithmic correlation. The
correlation coefficients (coef) and the standard deviations (s) are indicated for each regression line.
After [36].

fractal dimensions are common in various types of natural patterns or natural-pattern-based data sets [9]. They point to a complex structure of the pattern or data set and have been taken as an indication of different subsequent processes [9, 12].

In order to gain information about the anisotropy of the fracture pattern, the fractal dimensions (slopes of the regression lines) have been plotted in relation to the line orientation in so-called *fractal-dimension orientation diagrams* (*DOD*, [36]) that indicate, in general, the pattern anisotropy. A complete isotropy of the pattern would result in the same fractal dimension for all directions. In that case the data points would be oriented on a circle around the center of the diagram. However, the diagram shows a clear deviation from such a configuration. The data points approach an elliptical distribution (Fig. 5). The orientations of the long and short ellipse axes *a* and *b* represent the orientations of maximum and minimum fractal dimension. The axial ratio is named *azimuthal anisotropy of fractal dimension* (*AAD*, [36]).

6 Grain boundary patterns

The nature of crystalline material does not only depend on the physical and chemical characteristics of the different components of the material but also on the type of boundaries between the different crystals. These grain and phase boundaries affect the tensile and shear resistance of rocks — an important parameter in engineering geology — as well as the strength of artificial material after annealing, welding or other deformation and heating processes. They affect wave velocities or play a part for the permeability of the material in relation to fluids or gas and, consequently, for the reactivity of the material. Moreover, grain or phase boundaries provide information about the material history. This is important with respect to metamorphic rocks, of which the deformation and metamorphism events can be studied by analyzing grain and phase boundaries and their patterns.

Sutured (bulged) grain boundaries represent excellent examples of structures that are self-similar over several orders of magnitude, from the nanometer to the millimeter scale [12]. Generally, this self-similarity can be related to the fact that grain boundary suturing is caused by atomic diffusion across the boundaries [19]. It is known that diffusion of any kind of matter in any kind of host material may lead to self-similar structures [9]. Sutured

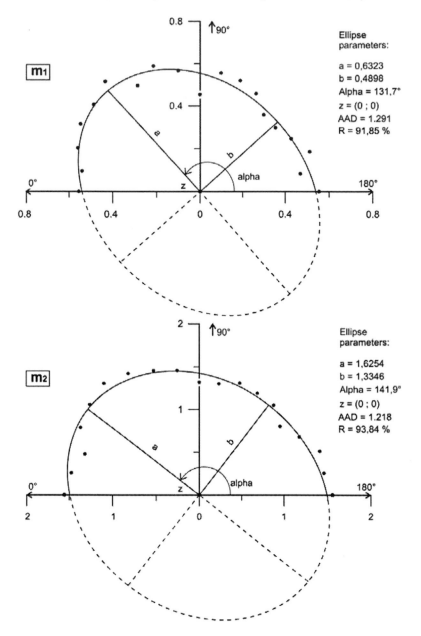

Figure 5: *Fractal-dimension orientation diagrams* (*DOD*) of the section shown in figure 3: In a non dimensional circular coordinate system, the slopes m_1 and m_2 of the regression lines, as exemplified in Fig. 4, are presented for 18 different directions in 10° steps from 0° to 180°. The 0°/180° line is shown in Fig. 3. The slope values are plotted as distances from the center (z) of the diagram towards the outside. For both diagrams the best-fit regression ellipse is shown, together with the long and short axes, *a* and *b*, and the inclination alpha of *a* versus the 0/180° line. R = correlation coefficient; *AAD = azimuthal anisotropy of fractal dimension* = ratio of *a* and *b*. After [36].

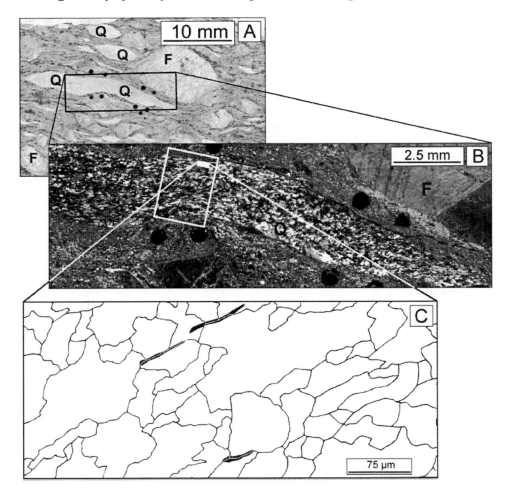

Figure 6: Photomicrographs [A,B] and line drawing [C] of the analyzed quartz-porphyric dyke from the Alpine „Root Zone" (Val Loana, Western Alps); sample KR2257. [A] Non deformed feldspar phenocrysts (F) and deformed lenticular quartz grains (Q) are aligned in a fine-grained well-foliated matrix. The box indicates the position of figure B. View of complete thin section; parallel polarizers. [B] Microfabric details of a deformed and dynamically recrystallized quartz lens. The large box indicates the analyzed entire area of the quartz lens, with the short box side parallel to the 0-180° line shown in figure 7A. The small white box marks the position of figure C. Crossed polarizers. [C] Line drawing of the grain boundary pattern. Black flakes represent biotite grains. The white lines indicate the probable grain boundary positions in the absence of biotite, as taken for the analysis.

grain boundaries can be measured under the polarizing microscope over about 1 to 2 orders of magnitude (from about 5 to 500 μm). They can serve as a geothermometer that provides deformation temperatures [12]. Other investigations show that the fractal dimension of sutured grain boundaries may also serve as a measure of the strain rate [31]. Both results are in agreement because the strain rate has the opposite effect on the formation of deformation microfabrics as the temperature.

The studies by [12] and [31] are dealing the grain boundary patterns as isotropic patterns and measure them by the box-counting and perimeter method, respectively. However, grain boundary patterns are anisotropic in many cases, mainly due to the preferred orientation of crystals in deformed and heated crystalline material, e.g. in metamorphic rocks or in welded and annealed metal work sheets. This grain boundary anisotropy leads to an anisotropy of material characteristics, as mentioned above: the tensile and shear resistance of rocks, the strength of artificial material after welding, the fluid percolation etc..

A modified perimeter method allows the quantification of the grain boundary pattern anisotropy by determining a direction-related fractal dimension, as exemplified by means of a grain boundary pattern of recrystallized quartz. An originally euhedral quartz crystal from a porphyric dyke of the Alpine „Root Zone" (Val Loana, Western Alps) was flattened to lenticular shape and recrystallized at temperatures of ~400-300°C during the late-Alpine deformation and retrograde Lepontine metamorphism (Fig. 6A,B) [10]. The recrystallized small new grains show a dimensional as well as a crystallographic preferred orientation. The dimensional orientation corresponds to a grain boundary orientation and represents an anisotropic grain boundary pattern (Fig. 6C). In order to record and quantify this anisotropy, a set of 5 parallel lines is superposed the grain boundary pattern and rotated in 10° steps (Fig. 7A). The accumulation of all the shortest grain boundary sections between two intersecting points forms the persistent grain boundary curve to be studied (Fig. 7B,C).

The analysis is performed by the normal perimeter method [9] and results in an average fractal dimension D for each direction. The altogether 18 values are plotted from a center towards the respective direction (Fig. 8). They show an elliptical distribution. The axial ratio of the best-fit ellipse, i.e. the *azimuthal anisotropy of fractal dimension* (*AAD*) is 1.14 and represents a quantification of the grain boundary pattern anisotropy. Since the variation of the D values is below the variation of the long and short ellipse axis, the elliptical distribution of the D values is statically significant. It needs to be emphasized that this type of anisotropy is only loosely (and inversely) related to the grain shape anisotropy as presented below.

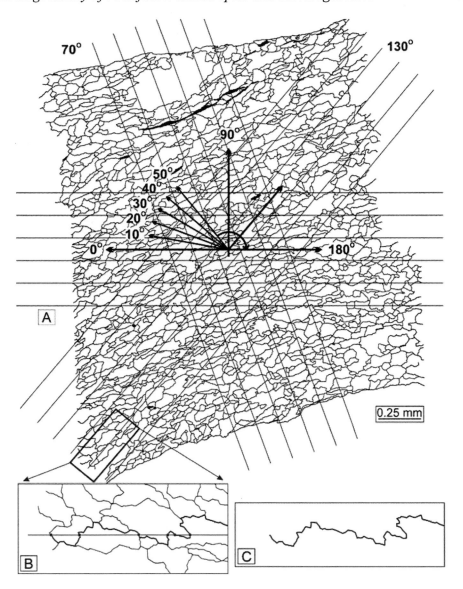

Figure 7: The modified perimeter method; line drawing of the analyzed total quartz grain boundary pattern, with secondary particles in black [A]. The pattern is overlaid by equal-spaced parallel lines, rotated in 10° steps exemplified by two sets of lines in 70° and 130° position. Each line leads to a se-quence of intersection points [B]. They form the support points of the grain boundary curve that is constructed as accumulation of all the shortest grain boundary sections between the points. The end points are located on the boundaries of the two last complete grains at each end of the line. The analysis of each grain boundary [C] is performed by the normal perimeter method [9] and results in a double-logarithmic plot of the step length versus the total polygon length. The slope of the linear re-gression gives the fractal dimension D, namely one average value from the five curves of each direction.

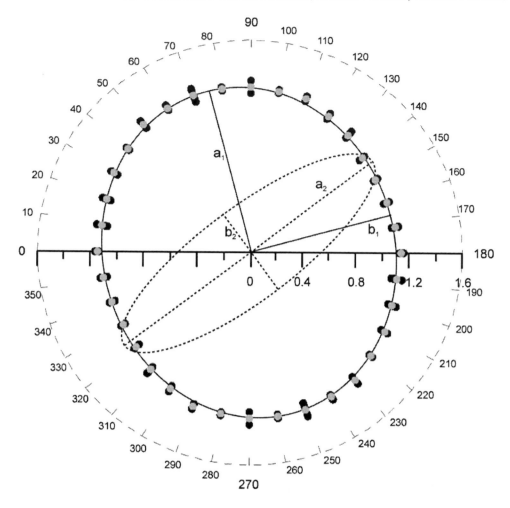

Figure 8: *Fractal-dimension orientation diagrams (DOD)* of the grain boundary pattern shown in figure 7A. In a non dimensional circular coordinate system, the fractal dimension of the modified perimeter method is presented for 18 different directions in 10° steps from 0° to 180°. The 0°-180° line is shown in Fig. 6B. The D-values are plotted as distances from the center of the diagram towards the outside (gray circles = average; filled circles = total range of D values). The best-fit regression ellipse is shown, together with the long and short axes, *a1* and *b1*. Axial ratio of the best-fit ellipse, i.e. *azimuthal anisotropy of fractal dimension (AAD)* = 1.14. The variation of the D values is below the variation of the long and short ellipse axis. The average dimensional shape of the grains, calculated on the basis of 625 grains with axial ratio > 2.39 (= average elliptical fit of all 1473 analyzed grains), is represented by a best-fit ellipse (short-broken line) with an axial ratio of the long and short axes *a2* and *b2* of 3.33.

In order to determine the grain shape anisotropy, the best-fit ellipse of the grain shapes with an axial ratio > 2.39 has been plotted (Fig. 8). It shows an ellipticity of 3.33. The long ellipse axis roughly represents the intersection line of the local flattening plane with the thin section. This direction is oriented at a high angle, but not perpendicular, to the long axis of the best-fit ellipse of the D values. The shapes of the two ellipses are clearly different, i.e. the azimuthal anisotropy of fractal dimension (AAD) is much weaker than the dimensional anisotropy of the grains. This is still valid if all grains are taken instead of only grains with an axial ratio > 2.39. We interpret this difference as the result of a strong ductile flattening versus a more isotropic grain boundary migration during and after deformation and/or as an only weak effect of the crystallographic preferred orientation on the grain boundary migration.

The quantification of the grain boundary anisotropy, based on the AAD value, allows the comparison of grain boundary patterns with different anisotropies and, last but not least, the correlation with parameters of grain boundary formation, such as temperature, strain rate, differential stress and different material properties.

7 Discussion

The physical and partly chemical properties of rocks or artificial crystalline material (particularly the tensile and shear resistance of rocks, the reactivity and the permeability of the material in relation to fluids, the strength of artificial material after annealing, welding or other deformation and heating processes) are strongly dependent on the material fabric. Since approximately two decades, fractal geometry offers methods for the quantification, mainly in 2-d, of self-similar or self-affine fabric patterns. They lead to a single value, the fractal dimension that characterizes the entire pattern of a studied area. However, such a quantification disregards that many fabric patterns are inhomogeneous, due to local material differences or variations of the stress field etc., as well as anisotropic, due to shear stress or flow etc., even in small areas. New modifications of fractal geometry techniques are designed to determine the inhomogeneity as well as anisotropy of fabric patterns. Three of these techniques are presented: (i) the 'map-counting' method for the quantification of pattern inhomogeneities, (ii) a modified Cantor-dust method for the quantification of anisotropies of fracture networks, and (iii) a combined Cantor-dust and perimeter method for the quantification of grain boundary patterns.

The map-counting method is a variation of the box-counting method, based on a kriging procedure. This method was established to analyze crystal distributions in a porphyric granite, however, can be applied in principle to different patterns, for example also to fracture patterns in rocks or artificial material. This method requires a sufficiently large area of a sufficiently finely structured pattern. It leads to an isoline map of the fractal dimension, which reflects the variation of the 'fractality' of the pattern. The method can be easily automated but, as a disadvantage, the resolution is rather low.

The application of modified Cantor-dust and perimeter methods to the quantification of pattern anisotropies has been demonstrated for grain boundary patterns, i.e. patterns on the micro scale, and for fracture patterns, i.e. patterns on the millimeter to centimeter scale. In both cases the directional distribution of the fractal dimension is elliptical. The deviation of the data points from a circular arrangement indicates the anisotropy of the grain boundary as well as of the fracture patterns. The ellipticity, i.e. the ratio of the long and short principal axes, represents the so-called azimuthal anisotropy of the fractal dimension D (AAD, [36]). The AAD value quantifies the pattern anisotropy and, consequently, allows the comparison between the anisotropies of different patterns. In the case of the grain boundary pattern, the difference between its AAD value and the ellipticity of the dimensional orientation of the grains throws light on the relationship between grain boundary migration and the development of the crystallographic preferred orientation and on the different periods of their formation.

In the case of the fracture pattern, the elliptical arrangement of the data points reflects shearing in 2-d, in agreement with the fact that the pattern derives from a brittle shear zone. Locally the data points clearly deviate from an elliptical arrangement. The reason is still unknown. However, it may be expected that the internal structure of such data point arrangements comprises additional information on specific pattern characteristics and, furthermore, on more sophisticated details of the pattern-forming processes, as postulated in case of grain boundary patterns [12, 30]. In order to investigate such data point structures in a better way, smaller steps of the angular variation, e.g. 1-2°, are required. Based on the partly already installed, but in any way possible, automatization, such data intensive investigation should be possible without great deal of time.

In general, the presented methods are first steps towards a quantification of pattern inhomogeneities and anisotropies. Such quantification provides more detailed information about patterns and facilitates the comparison between the anisotropies and inhomogeneities of different patterns and parts of patterns. In addition, the quantification provides a basis for comparing naturally, technically and experimentally produced fabrics. Pattern quantification is also expected to form a future useful fundament for the analysis of fabric-forming processes and of the interactions between such processes and the structured material. A better knowledge about fabric-forming processes is important for many fields of the geo- as well as engineering sciences.

Acknowledgements

Thanks are due to Dimitrios Kolymbas for organizing the Exploratory Workshop on Fractals and Geotechnical Engineering at the University of Innsbruck/Austria and providing such a great opportunity for exchange of information and new ideas in this field of science. The studies, on which our contribution is based, have been financially supported by the Technische Universität München and privately by the authors.

Bibliography

[1] Barton, C.C.: Fractal analysis of scaling and spatial clustering of fractures. In: Barton, C.C., La Pointe, P.R. (Eds.), Fractals in the Earth Sciences. Plenum Press, New York, pp. 141-178, 1995.

[2] Blenkinsop, T.G.: Fracture spacing distributions in rock. International Symposium on Fractals and Dynamic Systems in Geoscience, Book of Abstracts, Johann Wolfgang Goethe Universität, Frankfurt am Main/Germany, 6-7, 1993.

[3] Dubois, J., Cheminée, J.-L.: Application d'une analyse fractale à l'étude des cycles éruptifs du Piton de la Fournaise (La Réunion): modèle d'une poussière de Cantor. C.R. Acad. Sci. Paris, 307, pp. 1723-1729, 1988. Cited in: Velde et al., *Tectonophysics*, 179, pp. 345-352, 1990.

[4] Feder, J.: Fractals. Plenum Press, New York, pp. 283, 1988.

[5] Gillespie, P.A., Howard, C.B., Walsh, J.J., Watterson, J.: Measurement and characterization of spatial distributions of fractures. *Tectonophysics* 226, pp. 113-141, 1993.

[6] Harris, C., Franssen, R., Loosveld, R.: Fractal analysis of fractures in rocks: the Cantor's Dust method - comment. *Tectonophysics*, 198, pp. 107-111, 1991.

[7] Hirata, T.: Fractal Dimension of Fault Systems in Japan: Fractal Structure in Rock Fracture Geometry at Various Scales. *Pure and Applied Geophysics* 131, pp. 157-169, 1989.

[8] Jaeger, J.C., Cook, N.G.W.: Fundamentals of Rock Mechanics. Methuen & Co., London, pp. 513, 1969.

[9] Kaye, B.H.: A random walk through fractal dimensions. VCH, Weinhein, pp. 421, 1989.

[10] Kruhl, J.H.: Textures and c-axis orientations of deformed quartz crystals from porphyric dikes of the Alpine 'Root Zone' (Western Alps). *Geologische Rundschau* 75, pp. 601-623, 1986.

[11] Kruhl, J.H.: The formation of extensional veins: an application of the Cantor-dust model. In: Kruhl, J.H. (ed.), Fractals and Dynamics Systems in Geoscience, Springer, Berlin / Heidelberg, pp. 95-104, 1994.

[12] Kruhl, J.H., Nega, M.: The fractal shape of sutured quartz grain boundaries: application as a geothermometer. *Geologische Rundschau* 85, pp. 38-43, 1996.

[13] Mandelbrot, B.B.: The Fractal Geometry of Nature. Freeman, San Francisco, 1983.

[14] Manning, C.E.: Fractal clustering of metamorphic veins. *Geology 22,* pp. 335-338, 1994.

[15] Marone, C., Scholz, C.H.: Particle-size distribution and microstructures within simulated fault gouge. *Journal of Structural Geology* 11, pp. 799-814, 1989.

[16] Nagahama, H.: Fracturing in the solid earth. Science Report Tohoku University, 2nd ser. 61, pp. 103-126, 1991.

[17] Nagahama, H.: Fractal fragment size distribution for brittle rocks. *International Journal of Rock Mechanics and Mining Sciences & Geomechanics*, Abstracts 30, pp. 469-471, 1993.

[18] Orford, J.D., Whalley, W.B.: The use of the fractal dimension to quantify the morphology of irregular-shaped particles. *Sedimentology* 30, pp. 655-668, 1983.

[19] Passchier, C.W., Trouw, R.A.J.: Microtectonics. Springer, Berlin / Heidelberg, pp. 289, 1996.

[20] Peitgen, H.-O., Saupe, D. (eds.): The Science of Fractal Images. Springer, NewYork / Berlin / Heidelberg, pp. 312, 1988.

[21] Peitgen, H.-O, Jürgens, H., Saupe, D.: Fractals for the classroom, Part 1. Springer, New York, 1992.

[22] Peitgen, H.-O, Jürgens, H., Saupe, D.: Bausteine des Chaos: Fraktale. Rowohlt, Reinbek, pp. 514, 1998.

[23] Peternell, M., Andries, F., Kruhl, J.H.: Magmatic flow-pattern anisotropies - analyzed on the basis of a new 'map-counting' fractal geometry method. DRT Tectonics conference, St. Malo, Book of Abstracts, 2003.

[24] Pettijohn, E.J.: Sedimentary Rocks. 3rd edition, Harper & Row, New York, pp. 628, 1975.

[25] Sammis, C.G., Osborne, R.H., Anderson, J.L., Banerdt, M., White, P.: Self-Similar Cataclasis in the Formation of Fault Gouge. *Pure and Applied Geophysics* 124, pp. 53-78, 1986.

[26] Sammis, C., King, G., Biegel, R.: The Kinematics of Gouge Deformation. *Pure and Applied Geophysics* 125, pp. 777-812, 1987.

[27] Simpson, G.D.H.: Synmetamorphic vein spacing distributions: characterization and origin of a distribution of veins from NW Sardinia, Italy. *Journal of Structural Geology* 22, pp. 335-348, 2000.

[28] Smalley, R.F., Chatelain, L.-L., Turcotte, D.L., Prevot, R.: A fractal approach to the clustering of earthquakes: applications to the seismicity of the New Hebrides. *Bulletin of the Seismological Society of America* 77, pp. 1368-1381, 1987.

[29] Suteanu, C., Zugravescu, D., Munteanu, F.: Fractal Approach of Structuring by Fragmentation. Birkhäuser, Basel, PAGEOPH 157, pp. 539-557, 2000.

[30] Suteanu, C., Kruhl, J.H.: Investigation of heterogeneous scaling intervals exemplified by sutured quartz grain boundaries. *Fractals* 10/4, pp. 435-449, 2002.

[31] Takahashi, M., Nagahama, H., Masuda, T., Fujimura, A.: Fractal analysis of experimentally, dynamically recrystallized quartz grains and its possible application as a strain rate meter. *Journal of Structural Geology* 20, pp. 269-275, 1998.

[32] Turcotte, D.L.: Fractals and Fragmentation. *Journal of Geophysical Research* 91, pp. 1921-1926, 1986.

[33] Turcotte, D.L.: Fractals in Geology and Geophysics. *PAGEOPH* 131, pp. 171-196, 1989.

[34] Velde, B., Dubois, J., Touchard, G., Badri, A.: Fractal analysis of fractures in rocks: the Cantor's Dust method. *Tectonophysics* 179, pp. 345-352, 1990.

[35] Velde, B., Dubois, J., Moore, D., Touchard, G.: Fractal patterns of fractures in granites. *Earth & Planetary Science Letters* 104, pp. 25-35, 1991.

[36] Volland, S., Kruhl, J.H.: Anisotropy quantification: the application of fractal geometry methods on tectonic fracture patterns of a Hercynian fault zone in NW-Sardinia. *Journal of Structural Geology* (in press), 2004.

Fractals and scale effects in fractured multi-layered red beds

Christian A. Hecht

Institut für Geologische Wissenschaften, Von-Senkendorff-Platz 3, D 06120 Halle Saale

e-mail: hecht@geologie.uni-halle.de

Abstract

This contribution is concerned with the occurrence and the geomechanical and hydraulic effects of fractals in multi-layered sequences, in brittle deformed red beds in particular. The observations comprise two common examples of natural fractals, one on the grain scale and one on the outcrop scale. A new concept for the description of the geomechanically relevant order of a structure is introduced and applied to the examples of the two fractals. On the basis of different ranges of the geomechanical order of a system at different scales, a theoretical scale transition model is proposed. The fractal phenomena and their potential for correlation purposes are discussed within the framework of geomechanical application. The discussion also includes more general aspects of the degree of order of a system and the role of fractals in genetic analysis and behaviour prediction of multi-layered red beds.

1 Introduction

Scale transitions and scale effects are among the most interesting and challenging problems in engineering geology and geotechnical sciences. The commonly asked questions in this field of research are how to relate rock strengths or hydraulic rock properties across scale boundaries in order to make large systems predictable and calculable. Answers to these questions are very difficult because there is an unbalance in the knowledge and experience that we gain from the laboratory scale compared to the scales at which we interact with geological systems. The application of methods of fractal geometry in engineering geology is not very popular and there is certainly a lot more to explore. Fractals are infinite geometrical objects characterized by properties like self-similarity, self affinity and scale invariance. We know that geological objects can look very similar at different scales and we are not able to infer their absolute size without any scale information. However, geological objects are rarely self-similar in a mathematical sense and even in those cases where they statistically are, they cover only a few orders of magnitude.

Engineering geologists need to express rock properties precisely by numbers. The petrophysical rock properties cause us less problems because we can get reasonable numbers from rock experiments. As a matter of technological limits, the size of experiments is limited and our empirical knowledge diminishes rapidly with increasing scales. Another important question is how to precisely describe the structure and composition of rocks by numbers. The methods of fractal analysis appear very interesting in both aspects because they cover different scales and allow us to characterize the structure or irregularity of an object by its fractal dimension.

The scope of this paper is to delineate the potentials and limits of fractal analysis for the solution of problems of rock characterisation and scale effects in engineering geology. The study is mainly concerned with layered coarse grained clastic rocks. It presents a guide line for the analysis of complex systems on different scales using a function of structure and composition and presents a theoretical scale transition model based on that concept. The application of fractals is discussed within the framework of the presented models.

2 Observations

2.1 Application of fractal analysis

Since the publication of the book "The Fractal Nature" by Mandelbrot [9] fractal methods have been widely applied to geological structures. The applicability and wealth of fractal methods has been neglected by many scientists perhaps because in the years when fractals were "en vogue" many examples of misunderstanding and misapplication were published. Today fractal methods are well confined and their applicability for example in material sciences is unquestioned. Although we know that many if not most geological structures are not fractal "sensu stricto", the concepts of fractal geometry have conducted our minds to new principles like hierarchy, self-similarity and scale invariance of geological structures and processes [3]. Fractal principles also deeply support our thoughts about self-organizing phenomena of geological systems. One of the most popular examples of fractal behaviour in geology is fragmentation [10]. An example of fractal fragmentation of a sandstone layer through strike slip faulting is shown in Figure 1 [3] [5].

Figure 1: left: fractal fragmentation of a sandstone layer (layer thickness approx. 1.5 m)
right: Fractal fragmentation model (after Turcotte 1999)

Another example for the applicability of fractal methods is the analysis of the grain size distribution of dense granular packing types [4]. The well known fractal Appolonian packing is a good two dimensional disc model for hierarchically packed grain assemblies. An example of the Appolonian packing is shown in the next Figure (Fig. 2). The Appolonian packing is not a self-similar but a self-invers fractal, with very complex interstitial space geometries. It is one of the most universal two dimensional disc models, because it allows the use of different radii from the beginning of the packing process. It is a very good analogue for poly-phase sedimentary rocks and for mix in place mineral assemblages, where first a coarse grain framework is built up, that is later filled by sand slurry.

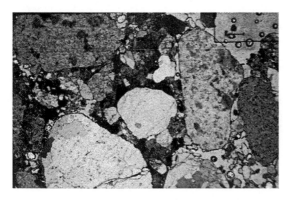 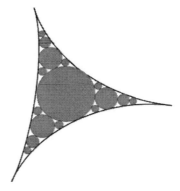

Figure 2: left: Densely packed fine grained conglomerate (length of thin section approx. 2 cm)
right: Model of Appolonian Packing (after Mandelbrot 1977)

From the theories of fractal analysis it follows, that in both examples only the geometries of the structures are analysed by the determination of the fractal dimensions D of the grain or fracture patterns. The disadvantage of the methods is that the dimension number tells us nothing about the properties of the single objects that compose a fractal pattern. In the example of fractal fragmentation (Fig. 1) the small fragments in the centre of the fault may be stronger weathered and therefore mechanically weaker than the bigger blocks farther away from the fault zone. In the example of Appolonian packing (Fig. 2) the single grains may have different shapes and strengths and therefore different mechanical properties. These findings indicate that, from a geomechanical point of view we need to go beyond the analysis of the structure of a system and also carefully look at the composition of its elements (Fig. 3, Fig. 4).

2.2 The concept of geomechanical order

The following section introduces a guideline for the determination of the geomechanically relevant degree of order of a system as a function of structural order and compositional order. The first part, which is called "structural order", is determined at a certain scale by distribution statistics of the elements of a system and by the determination of the masses, shapes and sizes of the individual elements. Beyond the pure structures and geometries, the degree of order of a system further depends on differences of the petrophysical properties of its elements. This second part is the "compositional order" of the system. Finally, the total degree of order, the "geomechanical order" of a structure is a function of the geomechanically relevant geometries (structural order) and material properties (compositional order). To give a simple illustrative example, we consider two sandstones that have identical grain size distribution curves, in other words that have the same structural degree of order. We assume that sandstone No. 1 comprises 95 % of quartz grains and 5% of feldspars and sandstone No. 2 70 % of quartz grains and 30 % of feldspars. Because quartz grains and feldspars have differing single grain properties, for example shape and strength, sandstone No. 2 has a lower degree of compositional order because it comprises a higher percentage of weaker elements than sandstone No. 1. This principle applies to many structures at different scales (Table 1). The calculation of the degree of order of a system is certainly not trivial, because it requires a selection of the right parameters at individual scales. However, once the parameters are elaborated and normalized for example to percentage of order scale transition becomes possible.

TYPE OF ORDER	SAMPLE SCALE	OUTCROP SCALE	CRUSTAL SCALE
Structural order	Grain size distribution	Fracture patterns	Fault patterns
	Packing density	Multi-layer geometries	Basin geometries
	Grain contacts		
	Coordination numbers		
Compositional order	Grain composition	Fracture properties	Fault properties
	Grain properties	Multi-layer properties	Basin fill properties
Geomechanical Order	Fabric behaviour	Fracture propagation	Fault kinematics, Basin
		Fracture connectivity	development

Table 1: Summary of multi-scale types of order in engineering geology.

2.3 Order on the sample scale

Grain size distribution curves of soft rocks or mineral mixtures cover a wide variety from single size mixtures to multi size mixtures [7]. Only a small number of them is hierarchically packed, which results in fractal grain size distribution curves [4]. These rocks or materials generally have extraordinary petrophysical and geomechanical properties like high density and strength and low porosity and permeability. The occurrence of single beds with fractal grain size distributions can seriously disturb the hydraulic connectivity of a sequence by acting as a barrier for fluids and hydrocarbons. Fractal single layers can also influence the geomechanical behaviour by being very stiff, brittle and highly abrasive. It can be shown that the type of packing is of prior importance for the density and porosity of a rock, but the strength also strongly depends on the composition of grains within a rock (Fig 3).

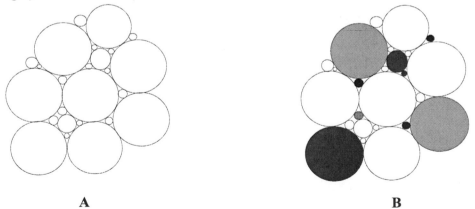

A   **B**

Figure 3: Models of Appolonian grain Packing. A: equal grain composition. B: unequal grain composition.

If we consider identical grain size distributions and cements, the strength of a rock would be a function of the content and the spatial distribution of weak grains. In the case of a fractal grain size composition we can use the dimension number for correlation purposes, if all grains have equal single grain properties (Fig 3 A). If the single grain properties, for example strengths, are different (Fig. 3 B) the fractal dimension D needs to be diminished for example by a constant of composition that accounts to the number and distribution of weak grains in the assemblage.

The equation would then read:

$$D_{eff} = D - C_{comp} \qquad (1)$$

D_{eff} = the geomechanically relevant dimension

D = fractal dimension

C_{comp} = Constant of composition and distribution of different grains.

The fractal dimension number of a grain assemblage can be calculated from log-log plots of the cumulative grain size distribution curve [4]. The question of how to precisely determine the constant of composition cannot be answered universally. It depends on the spatial distribution of grains of different strength and the range of single grain strengths variation within a certain assemblage. The spatial distribution of different grains can be obtained by thin section image analysis. The single grain strength is certainly not so easy to measure, but at least relative strength values can be estimated from general knowledge. It is also considered important to include a grain shape factor into the constant of composition if different grain shapes are present.

2.4   Order on the outcrop scale

The most interesting structures on the outcrop scale are fracture patterns. Fractal distributions of fracture patterns are calculated from line length counting or from box counting of images and trace maps [1] [2]. Again, only a small number of fracture sets is fractal and the measurements of the fractal dimension number give us just an estimate for the geometrical pattern. In red beds of Permocarboniferous to Triassic age fractal distributions on small to medium scale strike-slip faults were observed at different locations in Germany and the Southern Alps (Fig. 4). Fault zones of this type were found to be highly erosive and permeable, which is very important for underground operations. In most cases there is a decrease in the strength of the fragments towards the centre of the fault. The schematic illustration shows the principle model of fractal fragmentation (Fig. 4 A) and the geomechanical model, that has the same fractal pattern, but includes the observations of strength decrease towards the fault centre (Fig. 4 B).

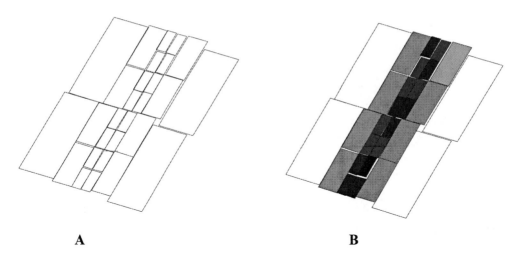

A **B**

Figure 4: Models of fractal fragmentation. A: equal fragment properties. B: unequal fragment properties.

2.5 Scale effects of fractals

Good examples for natural fractals like the Appolonian packing and the fragmentation patterns of local and regional faults only occur locally in multi-layered coarse grained sequences like red beds. Concerning scale effects, two remarkable groups of observations can be made. The first concerns the effects of local fractals within a sequence, while the second relates to the fractal geometry of large scale systems.

The occurrence of local fractal patterns has significant influence on the geomechanical and hydraulic behaviour of a sequence. A common example for local fractals are grain supported conglomerates with fractal grain size distributions, which are often found to be mechanically the strongest layers and at the same time the most impermeable ones. These fractal layers have a strong influence on the workability of red bed sequences in underground operations on one hand and on the hydraulic characterisation on the other hand. Other common examples are fractal faults in jointed rock masses, which generally create zones of weakness and high permeability. Although their distribution is not as frequent and as regular as that of joints, the mechanical and hydraulic influence of fractal fault zones is again very important. They certainly are good candidates for severe water inflows into underground operation sites.

The fractal geometry of large scale patterns for example of the multi-scale fracture set within a rock unit is strongly dependent on the analytical method [1] and on the point of view from which images are obtained [6]. In fact, there is a big difference whether one interprets a fracture pattern of photo lineations derived from a satellite image or a fracture pattern derived from a seismic section. Physical properties like pressure and temperature and lithologies do change on the vertical plane of observation, which inhibits self-similarity on large scales. Therefore fractal patterns occur only locally in certain layers or rock portions. In addition, long geological time processes are conserved on the vertical planes of observation, while on horizontal planes we look at synchronous processes. As a matter of more similarity of the external forces and the systems internal parameters, fractal patterns are more likely to occur in horizontal planes of observation. There are a lot of examples of natural fractals like river systems, fault systems and others on the earth surface [8], but hardly any on vertical geological cross sections [6]. For horizontally layered Permocarboniferous red beds the comparison of the distribution and geometry of fracture sets in different orientations has shown that selfsimilarity occurs on images horizontal to the earth surface (Fig. 5).

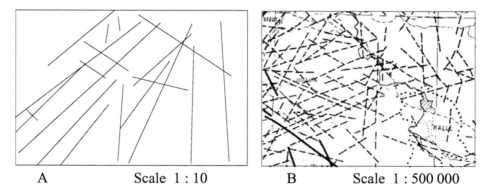

A   Scale 1 : 10   B   Scale 1 : 500 000

Figure 5: Selfsimilar trace maps of fracture patterns A: Horizontal image from the surface of a sandstone layer, B: Map of photo-lineation from the same region. Both images have equal orientations.

2.6 Scale transition model

Many empirical studies show that there are significant differences in strength or permeability values between rock samples and fractured rock masses or even larger units of the brittle earth crust. While the strength commonly decreases in an up-scaling procedure, the permeability often increases. The more different rock types are encountered in a multi layered sequence, the more arbitrary become the attempts to find a simple solution to the problem of up-scaling.

In view of this, the challenging question is: "How does the composition of a rock influence or determine the geomechanical behaviour of a single bed, how does the single bed behaviour influence the behaviour of a multi-layered rock sequence, how does the behaviour of a rock sequence influence the development of a large geological system and so forth?" A theoretical solution is proposed in Figure 6.

The model suggests that any large system reaches a certain percentage of degree of order (right point on the graph). Within a large scale system, for example a sedimentary basin, single samples can be analysed each displaying a certain percentage of order. In the graph three samples are displayed (left points on the graph). In an up-scaling procedure one has to move the values from small scales (left points) to large scales and all the traces must end in one point (right single point) which marks the order of the whole system. Doing so the transition passes through the medium scales of tens to hundreds of meters, which are the interesting scales at which engineering geology and geotechnical action takes place.

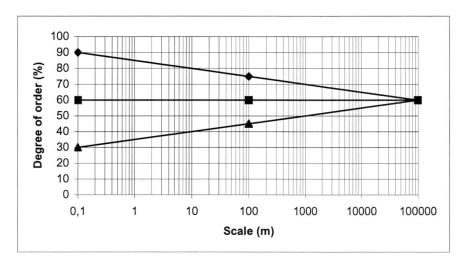

Figure 6: Scale transition model for the geomechanical order of sedimentary rock units. Points on the left correspond to the sample scale, points in the middle correspond to the outcrop scale and point on the right correspond the crustal scale.

In a system with a high degree of order, for example a sequence of sandstone layers of similar composition that was subjected to one phase of deformation, for example extension, the system's order is similar to the order at the medium and small scales. This leads to clustering of the points at small scales somewhere at high percentages of the degree of order, and the system has a small variance on the medium scales.

In a system with a low degree of order, for example a sequence of layers with strongly differing lithologies, the points on the sample scale have a high variation from low percentages of order to high percentages of order. As a result the systems order is generally lower and the variation of the degree of order on the medium scale is much larger. That is exactly what we observe on the medium scales, for example if we compare distribution statistics of joint patterns in sequences with equal single layer properties to sequences with different single layer properties.

2.7 Discussion

From a mathematical point of view, there are no scale limitations and no distinct working scales. Fractals are defined by geometrical attributes like hierarchy, self-similarity, scale invariance and so forth. Natural systems are not strictly fractal in a mathematical sense and the idea that fractals are everywhere is a sophism. There are good examples of fractals like, coastlines, river systems or fault zones, and less good examples like joint sets and others. In this study two natural fractals, the fractal strike slip faults and the Appolonian packing of dense granular materials, were introduced. A more general phenomenon of scale invariance of red beds is the fact, that they behave brittle on many scales. This does not necessarily mean that fractures are self-similar across many scales, which indeed is not the case and depends on the image orientation in this particular case.

From a geomechanical or hydraulic point of view we are farther interested in fracture properties like roughness, connectivity and others, which we can not obtain from pattern analysis only. Consequently, geomechanically relevant description models and combined models of fracture models and percolation models are necessary for new progress in this field.

The concept of geomechanical order accounts for the fact that the petrophysical condition and the geomechanical behaviour of a rock unit is the result of its structural and compositional features and its strain history. A system with strongly differing elements has a lower degree of order than a system that comprises similar elements. It can be proposed that the geomechanical and hydraulic behaviour of a system will be the more complex the lower the degree of order of the system is at the starting point. Magnifying a sedimentary system from the scale of samples to the scale of outcrops one needs to consider the vertical dimensions for example bed thickness, bedding relations and material changes and the square dimensions for example of ripple or dune shapes, of channel architectures and so forth. The results obtained with such an approach can be very different. If rock parameters and environmental parameters are uniform, the system is not sensitive to scale transitions over a large order of magnitudes, in other words the system is scale invariant in a broad sense.

Fractal patterns or fractal statistics are closely related with chaotic behaviour of a system. Rocks with a fractal grain-size distribution indicate chaotic processes of deposition for example through turbulent flow. The example of fractal fragmentation patterns of small-medium scale faults most likely represents a reactivation and overprinting of single joints through horizontal shearing. As the observations of this study show, fractals only occur locally under special circumstances. Apart from these particular examples, the majority of structures in red beds are distributed in normal, log normal or exponential ways, which leads to the conclusion, that processes involved with the development of red beds are rather deterministic.

This leads to the assumption that many initial conditions had a continuous influence at different steps of rock development, for example the initial properties of grain packing or bedding relations, which shows that rock formation processes of red-beds are fairly linear.

Conclusions

Natural fractals occur locally in multi-layered red beds. They are characterized by extraordinary geomechanical and hydraulic properties that can produce remarkable scale effects.

The applicability of fractal methods on large systems appears to be restricted to superficial patterns and processes

The principle of geomechanical order characterizes the geomechanically relevant parameters of a geological system as a function of structural order and compositional order. This principle also applies to fractal patterns.

The estimation of the degree of geomechanical order at different scales of observation allows scale transitions within large geological systems. They reveal interesting information of the complexity of a system on the scale of engineering geology and geotechnical interaction.

Bibliography

[1] Gillespie, P.A., Howard, C.B., Walsh, J.J. & Watterson, J.: Measurement and characterisation of spatial distributions of fractures. *Tectonophysics* 226: 113-141, 1993

[2] Harris, C., Franssen, R. & Loosveld, R.: Fractal analysis of fractures in rocks: the Cantor's Dust method-comment. *Tectonophysics* 198: 107-115, 1991

[3] Hecht, C. A. und Lempp, Ch.: Charakterisierung mechanischer Eigenschaften klastischer Locker- und Festgesteine mit Methoden der fraktalen Geometrie. Beiträge zur 12. Nationalen Tagung für Ingenieurgeologie, April 1999 in Halle/Saale,175-185, 1999

[4] Hecht, C. A.: Appolonian Packing and Fractal Shape of Grains improving Geomechanical Properties in Engineering Geology. *Pure and Applied Geophysics*, 157, 487 – 504, 2000

[5] Hecht, C. A.: Geomechanical and petrophysical properties of fracture systems in Permocarboniferous „red-beds" . Proceedings of the 38Th U.S. Rock Mechanics Symposium, DC Rocks, Washington, 1237-1245. Balkema Rotterdam, 2001

[6] Hecht, C.A.: Relations of self-similarity phenomena of multi-scale fracture systems to geomechanical and hydraulic properties of Permocarboniferous red beds. In: Benassi, A., Cohen, S., Istas, J. and Roux, D. (eds) Self Similarity and Applications. Annales de l'Universite de Clermont Ferrand, 2003

[7] Hecht, C. A.: Geomechanical models for Clastic Grain Packing. *Pure and Applied Geophysics*, 161, 331-349, 2004

[8] Korvin G.: Fractal Models in the Earth Sciences. Elsevier, 1992

[9] Mandelbrot, B.B.: The fractal geometry of Nature. Freeman: San Francisco, 1977

[10] Turcotte, D.L.: Fractals and Chaos in Geology and Geophysics. Cambridge University Press, 1999

Permeability of multiscale fracture networks

P. Davy[1], O. Bour[1], J.-R. de Dreuzy[1], C. Darcel[2]

1 Géosciences Rennes, université de Rennes I, campus de Beaulieu, Rennes, France

e-mail: Philippe.Davy@univ-rennes1.fr

e-mail: Olivier. Bour@univ-rennes1.fr

e-mail: Jean-Raynald.de-Dreuzy@univ-renne1.fr

2 ITASCA, 64 chem Mouilles, Ecully, France

e-mail: c.darcel@itasca.fr

Abstract

The paper aims at defining the flow models, including equivalent permeability, that are appropriate for multi-scale fracture networks. As a prerequisite of the flow analysis, we define the scaling nature of fracture networks that is likely quantified by power-law length distributions whose exponent fixes the contribution of large fractures versus small ones. Despite the absence of any characteristic length scale of the power-law model, the flow structure appears to contain three length scales at the very maximum: the connecting scale, the channeling scale, and the homogenization scale above which the equivalent permeability tends to a constant value. These scales, including their existence, depend on the fracture length distribution and on the transmissivity distribution per fracture. They are basic to define the flow properties of fracture networks.

1 Introduction

Modeling fluid transfers in fractured rocks still remains one of the main challenge of modern hydrology. Fractures are known to be key structures for the migration of hydrothermal fluids, but their spatial heterogeneity and complexity at all scales make difficult the definition of simplified, but relevant, flow models. A major difficulty comes from the multiple scales involved. Fractures occur at all scales, from micro-cracks which may be observed on thin sections up to pluri-kilometric faults which may break the entire crust. Beyond such a characteristic which implies a large distribution of fracture size, the geometry of fracture networks is also characterized by a wide distribution of orientations, of apertures and by a spatial repartition of fracture densities which may be inhomogeneous.

This complex geometry raises some fundamental issues about the hydraulic characterization and modeling of fractured media. The lack of any apparent characteristic scale for fracture network geometry does not make stand out the definition of a relevant scale that would help to define the basic properties of a relevant fluid flow model. The definition of a representative elementary volume which is basic to classical homogeneous models is in particular questionable in multi-scale heterogeneous systems. Even the definition of a pertinent scale, or scale range, to describe the geometry of a fracture networks, or to measure the hydraulic properties of the system, is not explicit.

In this paper, we give an insight into the way to deal with multi-scale fracture networks. We especially focus on the consequences of two basic statistical properties of fracture networks: (i) the fracture length distribution, and (ii) the fracture aperture distribution. Fracture length defines the spatial extent of the heterogeneity that a fracture may potentially induce on flow; fracture aperture defines the intensity of this flow heterogeneity by fixing fracture transmissivity. We demonstrate in the following sections that these two properties fully define the nature of flow and of the relationship between transport properties and fracture geometry. Other fracture network characteristics, such as fracture orientation distribution, do not change fundamentally these relationships, and can be easily incorporated in the presented theory with some simple adjustments of the basic model parameters.

We assume that the fracture length distribution is a power law. Beyond the relevance of this model to natural fracture networks – that will be discussed in the next section –, the power-law distribution model has the interesting property to have no characteristic length scale except its endmost limits that are the smallest fracture length l_{min}, and the largest one l_{max}. Studying the relationship between medium structure and fluid flow for such systems thus illustrates the type of difficulties that we may encounter when dealing with multi-scale heterogeneities. The issues of the definition of a relevant scale, or scale range, for fluid transport, and of the pertinent flow model are acutely addressed by such a distribution model. Although this case seems to be widely relevant in natural system, it is seldom studied in the literature (see however [1-4] about long-range correlated percolation in self-affine surfaces).

In this paper, we treat only systems where matrix is impervious, which corresponds to most of magmatic rocks. The end-member cases of infinitely long fractures, and of infinitesimally small cracks have already been fully characterized. Infinite fractures form parallel paths that can be treated analytically [5, 6]. The specific case of infinitesimally small fracture networks was extensively studied in the framework of the percolation theory [7]. We will show that these two end-member cases are encompassed in the power-law length distribution model, and that a much richer phenomenology appears when covering the range of all possible fracture distributions.

In the following, we first assess the validity of the power-law length distribution model for natural fracture systems. We then discuss the connectivity property, and eventually the network equivalent permeability as a function of the fracture network structure.

2 Multiscale geometry of fracture networks

Quantifying the scaling properties of fracture networks is crucial at least to assess the role of the different fracture families. It is really tempting to neglect unobserved fractures when calculating flow, arguing upon their small size. But small fractures are also numerous, and the balance between the large number of small structures and the small number of large ones is not *a priori* trivial. In this paragraph our concern is the quantitative determination of such scaling laws.

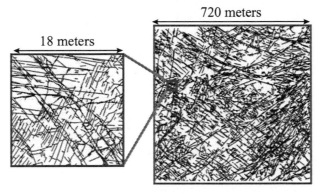

Fig. 1. 2D Fracture maps sampled from outcrops of the Hornelen basin.

The fact that fractures exist at all observable scales has been known since geology exists. The intriguing property is that similar patterns are observed at very different scales [8], emphasizing a kind of scale invariance of fracture networks (fig.1). Fractal concepts have shed new theoretical lights on such peculiar geometry [9], which emphasizes how deeply scales are linked in the fracturing process [10-13]. A pure fractal network has no intrinsic length scale, and the statistical distributions that describe object properties are power laws, the only functions that do not contain any characteristic length scale. The qualification and quantification of the basic power laws of fracture networks have been largely debated for the last fifteen years [14].

A fracture is defined by two basic scales, its length and the average distance to its neighbors, which defines in turn the fracture density. Power laws, if they exist, may apply on both the length distribution and the fracture density. We have proposed that the first-order statistics of fracture networks takes the following form:

$$N(l, L) = \gamma L^D (a-1) \cdot \frac{l^{-a}}{l_{min}^{-a+1}} \text{ for } l \in [l_{min}, l_{max}] \tag{1}$$

where $N(l, L)dl$ is the number of fractures having a length between l and $l+dl$ in a system of size L, γ is the fracture density term (number of fracture centers per fractal unit volume), a the exponent of the power-law length distribution, and D is the fractal dimension of fracture barycenters which fixes the scale-dependence of fracture numbers [15, 16]. The exponent a is really a quantitative measure of the balance between small and large fractures since it fixes the ratio between two fracture families of any length l and l', which is $(l'/l)^a$.

Validating this model and determining its parameters require a detailed description of fracture maps over a large range of scales, a condition which is generally not fulfilled by statistical analysis performed on single outcrop maps [14]. This difficulty can be overcome by analyzing fracture networks mapped at different scales of observation [17-21]. An example of such an analysis is given in figure 2 where seven two-dimensional fracture maps have been used to determine the underlying length distribution of fractures [15]. The comparison over scales requires knowing the fractal dimension D, which has been measured independently at about 1.8. Except for length below the resolution scales, data can all be mapped onto a single power law characterized by an exponent $-a$, with a of about 2.75 [15]. The fractal density term (α in equation 1) is equal to about 4.0. The consistency of the power-law statistical model over more than two decades has been tested with different methods [15]. It was made possible by the extraordinary dataset collected with different imaging techniques, from outcrop mapping to aerial photographs [21].

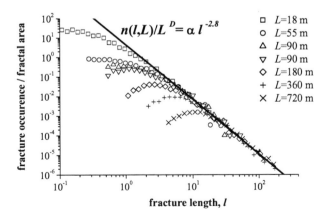

Fig. 2. Length distribution calculated for 7 fracture maps of the Hornelen basin (see fig.1)

The power-law length distribution model was found to describe a large variety of fracture networks, from small-scale joints (opening fractures) to large-scale faults (shear fractures) [21-26], and to fit with fracture growth models [16, 27]. The range of measured power-law exponents is quite large (see the compilation in [14]), but that it is not possible to discriminate between a statistical origin due to difficulties in achieving a robust statistical analysis, and some physical causes coming from the history of fault growth [27], or from the nature of fracturing [28]. In the example given in figure 1, which corresponds to a network of joints (tension fracture), we found that the power-law length exponent a and the fractal dimension D are about to fulfill the self-similar relationship: $a \approx D+1$ [15]. In contrast, we obtained a much smaller value of the length exponent a for the San Andreas fault system, with a about 2 [29].

Note that the power-law length model is still debated. Against such scaling theory is the evidence that intrinsic length scales exist in the physical system, which should control both processes and produced structures. The boundary conditions of the fracturing process and the layering of most geological formations are such length scales that may be identified in fracture distribution models. This addresses the issue of the validity domain of power-law models in terms of physical conditions, and/or of scale ranges. Note also that equation (1) is only a first-order description of fracture networks which gives the average number of fractures of a given length and at a given scale. It can be easily extended to incorporate other fracture properties such as aperture or orientation. But it does not describe possible correlations between the different geometrical parameters [29, 30].

Whatever the parameters, the power-law model predicts an occurrence of large fractures much larger than the one deduced from exponential or lognormal distributions that had been classically used. Since the fracture length is also a potential correlation distance for flow, this wide length distribution must entail some important consequences on the connectivity of fracture networks, and thus on the network equivalent permeability.

3 Connectivity of multi-scale fracture networks

In crystalline rocks where the matrix can be considered as impervious [31, 32], fracture network connectivity is a prerequisite to permeability. Except if fractures are large enough, the connectivity is ensured by the fracture clusters that span the whole system. In systems made up of small independent micro-cracks, the probability to have a spanning cluster (macro-connectivity) is closely related to the probability to connect two fractures (micro-connectivity). This micro-to-macro-upscaling was successfully described by the percolation theory [7], which statistically relates the global physical properties (connectivity, permeability, ...) to a density parameter of the global network, p, called the percolation parameter. There are several interesting concepts in percolation theory that help to understand how to deal with heterogeneity in systems close to threshold. The first concept is the correlation length ξ which is the typical size of the largest cluster in unconnected systems, or of the largest non-connected cluster in connected systems. ξ is really a correlation distance for flow: below the percolation threshold, it is the maximum distance for flow transfer; above it is the scale above which the system can be considered as homogeneous. The second concept is the percolation threshold defined as the point where the system goes from disconnected to connected by adding only one crack. This last added crack is among the "red links", which are essential to ensure connectivity, because, if one of them is removed from a connected cluster, the cluster becomes disconnected. In percolation theory, the percolation threshold is obtained at a fixed value of percolation parameter, $p=p_c$. As p tends to p_c, the correlation length increases to infinity – in practice, it cannot be larger than the system size–at a rate which depends only on the system dimension: $\xi \sim (p-p_c)^{-\nu}$ with $\nu=4/3$ in 2D and 0.88 in 3D. This divergence of the correlation length at percolation threshold due to the addition of a few cracks emphasizes the dramatic contribution of a few micro-cracks, making impossible the use of homogenization methods to predict the observed variations around connectivity threshold. The physical properties are also defined by the geometry of the structures that carry flow [7], mainly the infinite cluster, which macroscopically connects the system, and the backbone, which is the part of the infinite cluster that carries a non-nil flow (fig. 3). At the percolation threshold, the sample-spanning cluster, the backbone, and the red-links are fractal at any scale.

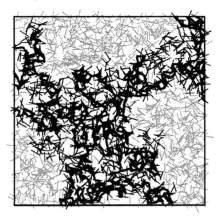

Fig. 3. Example of crack networks at percolation threshold. The connected cluster is in black and the backbone in bold.

In the early eighties, a great number of studies demonstrated that percolation theory applies to networks of small fractures such as micro-cracks. A major issue was the definition of the percolation parameter which depends on fracture orientations, fracture density and fracture lengths. The critical number of intersections per fracture and the critical density of cracks necessary to be at the percolation threshold have been established as functions of the orientation and length distributions [33].

In the nineties, the role played by a widely-scattered fracture length distribution was a major concern [34-36].

The extension of these studies to power-law length distributions, for which there is a significant number of fractures of length larger than the system size, was made both theoretically and numerically in 2D and 3D [37-39]. Except in [38], these results were obtained with fractures having a random orientation. From the results of previous studies, we indeed guess that a fracture orientation distribution modifies only slightly the expression of the percolation parameter but not the applicability of percolation theory as long as length and orientation are not correlated. We have first shown from intensive numerical simulations that, despite the wide range of length, there exists a percolation parameter p that fully describes system connectivity, which writes as :

$$p = N \cdot <e> \cdot \frac{<l_L^d>}{L^d} \qquad (2)$$

with l_L the fracture length which lies within the system of size L, N the total number of elements, d the Euclidean dimension of the system and e a form factor accounting for the fracture shape (equal to 1 in two-dimensional systems and to the eccentricity for ellipses in three-dimensional systems) [37-39]. The use of the included length l_L makes possible the extension to networks with fractures larger than the system size. Even when considering length and eccentricity distributions that cover several orders of magnitude, the percolation parameter at the percolation threshold, p_c, remains about constant with small variations mainly due to the eccentricity distribution.

Equation (2) makes possible a quantitative evaluation of the contribution of fracture lengths to connectivity. We showed for 2D networks that this contribution depends critically on the length exponent a defined by eq. 1 [37-39]. Thereafter we describe the main rules that may be simply extrapolated to 3D or fractal dimensions [38, 39].

- If a is very large, that is larger than 3 in 2D or 4 in 3D, the connectivity is controlled by fractures that are much smaller than the system size (fig. 4). The classical length distributions such as lognormal, exponential and gaussian, fall into this category since they decrease with length much faster than any power law. In that case, percolation theory fully applies with the same geometrical property and classical rules as the ones obtained on networks made up of fractures having all the same length much smaller than the system size, provided that the mean fracture length is correctly defined. As a consequence, multi-scale fracture networks can be advantageously replaced by fracture networks of homogeneous length.

- For *a* smaller than 2 in 2D and 3 in 3D, the fracture network connectivity is controlled by fractures having a size of the order or larger than the system size. The network is equivalent to a superposition of infinite fractures, with a connectivity rule which is simply the probability to encounter these large fractures. In other words, systems are always connected if they are larger than the average distance between two large fractures.

- For *a* between these two limits ([2, 3[in 2D and [2, 4[in 3D), the network connectivity is ruled by a combination of small and large fractures, the terms "small" and "large" being defined by comparison with the system size (fig. 4).

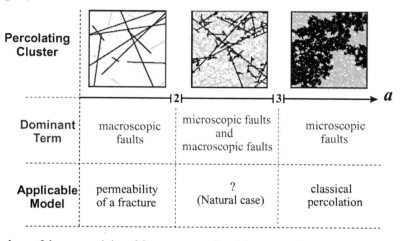

Fig.4. Typology of the connectivity of fracture networks with a power-law length distribution characterized by the exponent *a*

In case of fractal fracture networks for which D is lower than the Euclidean dimension, we expect the same rules to apply with a transition exponent of $a=D+1$ [28]. Note that $a<1$ implies that even the number of fracture is dominated by the largest fractures, a case which is obviously irrelevant for natural systems.

Although the percolation parameter p describes correctly connectivity properties whatever $a>1$, the classical percolation theory framework is fully relevant only for $a>D+1$. A basic assumption of this theory is that fractures are supposed to be much smaller than system size, at least for very large systems. This assumption is not valid if $1<a<D+1$ since the number of fractures larger than system size L increases as $N(>L)\sim L^{-a+D+1}$. As a very first consequence that illustrates the failure of percolation theory, the width of the transition (in terms of percolation parameter variations) from non-connected to connected networks remains large and does not vanish, even for infinitely large fracture systems. It implies that the percolation parameter at threshold depends not only on fracture density as in classical percolation theory, but also on the system size, L (fig. 5). In other words, for a given fracture system at fixed density, the network connectivity increases with scale so that it always becomes well connected at large scales. The critical scale L_c at which networks have a probability of 0.5 to be connected can be analytically calculated as a function of fracture parameters (density, power-law exponent a, …). In a bi-logarithmic density-scale diagram, L_c corresponds to a diagonal straight line whose slope depends on the length exponent a (fig. 5). For $L<L_c$, the fracture network is on average not connected although large fractures may be occasionally encountered in small systems. This peculiar scaling behavior has some practical and theoretical consequences on the network equivalent permeability that we develop below.

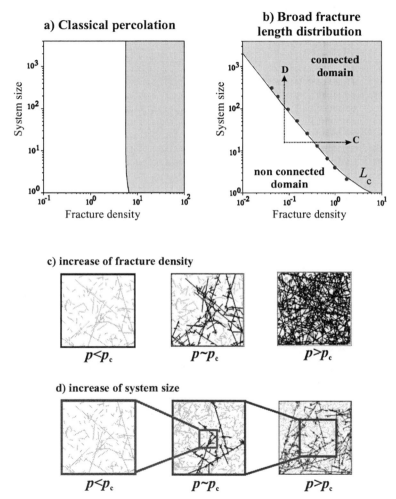

Fig. 5. System connectivity as a function of fracture density and system size in the cases of percolation theory (a) and of the multiscale fracture networks with $a<3$ (b). The zone where networks are connected is shaded in gray. (c) and (d) show the increase of permeability with density and scale respectively.

4 Network equivalent permeability

The large variability of geometrical and physical characteristics of fracture networks generates highly heterogeneous flow fields, for which classical modeling frameworks like homogenization may be irrelevant. Accounting for the heterogeneity may change the basic relations between the local medium properties, i.e. the fracture characteristics, and the global medium hydraulic properties. In this spirit, we focus on some key issues of both theoretical interest to understand the flow processes in fractured media and practical interest to design appropriate modeling frameworks for the network equivalent permeability:

- How connectivity influences permeability? What is the effect of the connection length and more generally of the geometrical network structure on permeability?
- Is there any scale-dependence of permeability like there is a scale-dependence of connectivity?
- What are the characteristic flow structures? Is there any evolution of the relevant flow pattern with scale? Is there any homogenization scale?

We address these issues from the main results obtained by [40-42] on 2D random synthetic networks with the following assumptions on fracture length and aperture distributions.

5 About fracture transmissivity

A fracture is an open void, or a zone of high permeability, which is characterized by a transmissivity value t_f, which is the permeability integral over fracture thickness (or aperture) [43]. For open fractures, this transmissivity is supposed to be related to the average fracture aperture to the power 3, by reference to the cubic law of two parallel planes [44-49]. The permeability of a unique fracture, whose width and length is much larger than the system size (but whose thickness is small and finite), thus decreases with scale as $k_f \sim t/L$ with t its transmissivity and L the system size (note that the permeability of a regular grid of infinite fractures has a constant permeability proportional to the transmissivity divided by the fracture spacing).

In addition to fracture length distribution (1) and scale, the probability distribution of fracture transmissivity $n(t_f)$ is a key input parameter in determining the network equivalent permeability. For open fractures, the local transmissivity distribution may be directly derived from the distribution of the average aperture per fracture. The few systematic measurements on fracture apertures seem to indicate that its distribution is a power law [50-52], entailing a power-law distribution of individual fracture transmissivity. However fractures are not only joints – i.e. open fractures – but also faults filled by a granular gouge, whose transmissivity is hardly determined except from hydraulic tests. Statistical analysis on such tests have, in general, led authors to propose a lognormal distribution for t_f [53-55], with a lognormal standard deviation b as large as 3 and a geometrical mean k_{mgm}. We thus consider both power law and lognormal distributions as potentially sound for transmissivity per fracture, with respectively main parameters b and t_{mg}, the lognormal standard deviation and the geometrical mean, and δ and t_{min}, the power-law exponent and the minimal fracture transmissivity.

The simultaneous modeling of the widely-scattered fracture length and fracture transmissivity distributions also raises the concern of possible correlations between these two parameters. If the longest fracture has the largest permeability, we expect a strong flow localization within these large structures, and thus a significant increase of the network permeability. The effect of a fracture length-fracture transmissivity correlation is also discussed in the following paragraphs.

6 Pertinent permeability models from scaling properties, and flow structure

The network equivalent permeability is fully defined by both the network structure, and local transmissivities. It is rather trivial to say that this twofold information is out of reach of the best geophysical method; in particular small-scale structures whose contribution on permeability is an important issue when trying to assess a relevant permeability model. The aim of the following sections is to discuss flow structure and permeability models, assuming that both fracture network geometry and fracture transmissivities belong to generic distribution models: power laws for fracture length distribution, and either power law or lognormal distribution for fracture transmissivity. The analytical description of permeability as a function of the geometrical and analytical characteristics is to be found explicitly in three previous articles [40-42]. In this paper, we discuss the main hydrologic models that come out from the range of admissible parameters for these distributions. We first discuss some basic properties that define general hydraulic models, i.e. permeability scaling and flow structure. Then we develop in the next 5 subparagraphs the main hydraulic models.

Note that the network permeability that we calculate is a statistical distribution, which is found to be remarkably well fitted by a lognormal function, whether the transmissivity distribution is lognormal or power-law. The distribution was inferred from extensive numerical simulations on 2D networks, practically from some hundreds of realizations for the largest networks containing up to some hundreds of thousands of fracture intersections, to some hundreds of thousands of realizations for the smallest networks.

Scaling is a key issue when attempting to assess the role of heterogeneities on flow. It reveals two basic features: the dimensionality of flow structure, and how network permeability samples local values. The former entails a decrease of network permeability with scale if the dimension of the structure that support flow is smaller than space dimension. The permeability of a single fracture for instance scales as L-1 with L is the system size. The latter effect (how permeability samples local values) scales as a consequence of the well-known statistical effects that the probability to encounter extreme values, large or small, increases with system size. Thus systems whose flow samples preferentially high- (low-, resp.) permeability domains are likely to have a positive (negative resp.) network-permeability scaling. This statistical effect fixes the scaling behavior of 1D, 2D and 3D homogeneous grids with a link-permeability distribution [56]. According to the different model parameters, permeability was found to have all the possible scaling behavior that is either decreasing with scale, constant with scale, increasing with a limit or increasing without a limit. The conditions that make the permeability model belonging to one of these four scaling relationships are explicated below.

The spatial distribution of flow is also a main characteristic of any flow model. Flow localization for instance is an expected consequence of spatial heterogeneity that reveals the capacity of fluid to select the path that minimizes viscous dissipation – i.e. that of largest permeability. In the following we use a simplified quantification of flow localization that amounts to comparing the main flow path – the path that carries the largest flow – with the others: we qualify a system as "channeled" or "distributed" whether the main flow path carries more flow that all the others together, or not. Note that the number of fractures in the main flow path is an important characteristic of the permeability structure.

The different hydraulic models that we describe below are defined from the style of permeability scaling (either decreasing, constant, or increasing), and of flow structure (extremely channeled or distributed). Note that a given network can belong to several hydraulic models depending on scale. Indeed we expect the flow structure to potentially contain three main length scales that may control permeability scaling:

- the connection length, i.e. the crossover scale at which some networks shift from disconnected to connected,
- the homogenization scale, at which the network equivalent permeability becomes constant (if it does),
- the channeling scale, above which the flow structure shifts from extremely channeled to distributed.

6.1 Percolation-like networks.

Percolation-like networks are encountered in 2D or 3D networks when the length distribution is largely dominated by fractures much smaller than system size [7, 40-42, 57]. For the power-law length distribution, this condition is fulfilled in 2D if the length exponent a is larger than 3 [40-42]. Indeed the probability of occurrence of a fracture of length larger than l, scaling as l^{a+1}, decreases always faster than l^2, so that fractures as large as the system size occur with a very low probability. Besides, the fracture length distribution has well-defined mean and standard deviation. Numerical simulations confirm that the fracture length distribution does not have any effect on the type of the hydraulic model. Another consequence is that the correlation between fracture length and fracture transmissivity is negligible.

There is a characteristic scale, the correlation length ξ, below which the network permeability decreases with scale and above which it is constant. Below the correlation length, flow is extremely channeled in the highest permeable path, called the critical path, and the network permeability is determined by its least permeable element (the critical bond). The network permeability is thus determined by a single element giving a simple method for estimating the network permeability. This method called the critical path analysis might be used for the upscaling of other phenomena [57, 58].

Above the correlation length, permeability reaches a limit and flow becomes homogenously distributed. The correlation length is practically the homogenization length and defines the Representative Elementary Volume (REV).

The permeability decrease and the correlation length depend on the transmissivity distribution. For the lognormal and log-uniform fracture transmissivity distributions, the permeability of the critical bond decreases and the correlation length increases when the fracture transmissivity distribution broadens [40, 57]. For a power-law fracture- transmissivity distribution with an exponent $-\delta$, the less permeable fractures are the most numerous (since its exponent δ is always considered to be larger than 1) and control always the network permeability. In this case the average permeability is fixed by the smallest fracture transmissivity; the rest of the distribution changes neither the network permeability nor the correlation length.

6.2 Unique-fracture networks

When a is small enough, that is smaller than 3 in 2D (or 4 in 3D), some networks below the percolation threshold are connected by a unique crossing fracture in which flow is completely channeled. Such large fractures occur with a probability proportional to $L^{\min(3-a,1)}$ according to eq.(1) [41] (note that this expression is obtained by considering all fractures in a space much larger than L since some fractures whose center is outside the system can cross it). For a smaller than 3 (in 2D), this type of connectivity that is ensured by a single fracture thus increases with the system scale L. That is why connectivity increases with scale. Since the permeability of a unique fracture decreases with scale as $1/L$ times the fracture transmissivity, the equivalent network permeability scales as $L^{\min(2-a,0)}$ if there is no scale effect on transmissivity – i.e. in the absence of correlation between fracture length and transmissivity. For $a<2$, permeability is constant because the increase of connectivity exactly offsets the decrease of the fracture permeability. For $3<a<2$, permeability decreases as L^{2-a}.

If there is a correlation between fracture length and transmissivity, the permeability scaling also includes the transmissivity scaling. For instance we can easily calculate this scaling in the case of perfect correlation with a power-law fracture-transmissivity distribution: $K \sim L^{\min(2-a+\beta,0)}$ with β the exponent of the relationship $t \sim l^{\beta}$ (because of the two power-law distributions β is equal to $\dfrac{a-1}{\delta-1}$). Since β is positive, the correlation effect tends to reduce the decrease of permeability with scale, and can even lead to a scale increase.

This regime prevails below the connectivity scale, i.e. below the scale at which networks shift from disconnected to connected.

6.3 Networks above threshold for which permeability is constant

In this case, at the percolation threshold, permeability becomes constant and flow becomes homogenously distributed in the network. This means that the connectivity, channeling and homogenization scales are all equal. This happens for constant fracture- transmissivity networks, i.e. networks in which all fractures have the same permeability (absence of fracture transmissivity distribution).

The correlation length for $a>3$ and the critical connectivity scale for $a\leq3$ play the same role of homogenization length. A network taken above this homogenization length is made up of a superstructure which looks like a grid whose mesh is itself made up of either a succession of links and blobs for $a>3$ or of a multiple-path multiple-segment structure for $a\leq3$ (fig. 6).

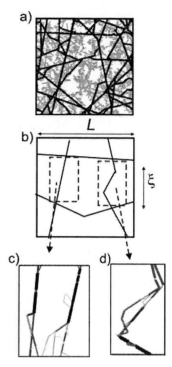

Figure 6. (a) Typical fracture network generated with *a*=2.5 and with a constant fracture transmissivity. Black fractures are from the super-structure of the backbone (b), whereas gray fractures can be put into the mesh element. (c) and (d) are sketches of the multipath-multisegment structure.

In the case where fracture length is "broadly" distributed (1<*a*<3), permeability increases with scale from the connectivity length to the channeling scale, at which flow becomes to be homogenously distributed in the network and where permeability has almost reached its limit. The channeling scale, at which extreme channeling vanishes, can be assimilated to the homogenization length, at which permeability stops to increase. It is not by chance that extreme channeling is linked to the permeability increase. In fact flow is extremely channeled in structures of increasing permeability with scale (fig. 7). When the extreme channeling vanishes, permeability stops increasing. Basically, when it exists, the channeling length is the average distance between two of the most permeable channels of the network.

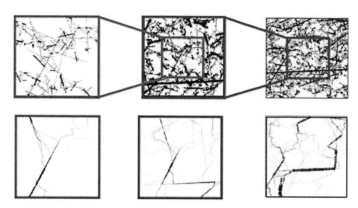

Figure 7: Illustration of the permeability increase for a given network generated for $a=2.5$ and a lognormal fracture transmissivity distribution not correlated with the fracture length distribution characterized by $b=2$. The network is analyzed at increasing scales from left to right ($L=50$, 100 and 200). The evolution of the flow pattern and more precisely the channeling in more permeable structures leads to an increase of the permeability.

This flow model, characterized by an extreme flow channeling and a scale-increase of permeability with a limit, occurs in four types of networks listed below. The network types differ by their generating parameters and by the amplitude of the permeability increase.

- Networks having a power-law length distribution with $1<a<3$, a lognormal permeability distribution, and no correlation between fracture transmissivity and length. The range of permeability increase only depends on the standard deviation of the fracture transmissivity distribution b and is equal to $\exp(b^2/2)$. The fracture length distribution determines the homogenization length. The broader the length distribution, the smaller the homogenization length and the faster permeability reaches its limiting value.
- Networks having a power-law length distribution such as $2<a<3$, a lognormal permeability distribution, and a perfect correlation between fracture length and fracture transmissivity, i.e. that the largest fracture is the most permeable one. The equivalent permeability increases with a limit. The amplitude of the permeability increase is much larger than in the case of the absence of correlation.

6.4 Networks above threshold for which permeability increases without a limit

In two cases of "broad" fracture length and permeability distributions, the equivalent permeability increases without a limit. The homogenization length is neither defined although the flow structure may be distributed in several paths.

The first case corresponds to a lognormal distribution for fracture transmissivity, a fracture length distribution whose characteristic exponent *a* lies in the range [1,2], and a perfect correlation between fracture transmissivity and length [40]. In this case, the equivalent network permeability diverges logarithmically. There is no homogenization length but there is still a channeling length at which extreme channeling vanishes.

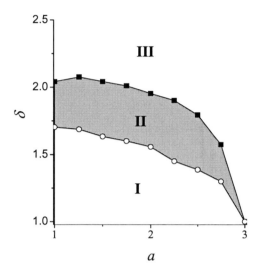

Figure. 8: Fracture length and permeability are power-law distributed with *a* and *δ* the characteristic exponent respectively. In zone I, flow is channeled in one path whatever the scale. In zone II, flow is first channeled for length scales smaller than the channeling length and becomes distributed in several different paths above. In zone III, flow becomes distributed above the connection length (connection length = channeling length).

The second case corresponds to a power-law fracture-transmissivity distribution, a fracture length distribution with a power-law exponent in the range [1,3]. The equivalent network permeability K increases as a power law with scale such as $K \sim L^\tau$ with τ given by [42]:

$$\tau = \frac{\left[\alpha_1 - \alpha_2 \cdot (a-1)\right] \min(3-a,1)}{\delta - 1} \qquad (3)$$

with α_1 and α_2 two constants close to 1 and 0.1. In this case, the fracture length distribution changes the permeability scaling. The permeability increase is steeper when *a* becomes closer to 1, i.e. when the system becomes mainly composed of large spanning fractures. Three possible cases have been found for the flow structure according to the characteristic exponents a and δ of the fracture length and permeability distributions:

- When a and δ larger are in zone III of figure 8 (roughly $\delta > 2$) flow becomes distributed at the percolation threshold, so that the connectivity and channeling lengths are equal.
- When a and δ larger are in zone III of figure 8, extreme channeling vanishes at a well-defined length scale larger than the connectivity length. Permeability still increases, whereas flow is not focused in a unique channel.
- Non-limited permeability increase and non-vanishing extreme channeling are obtained for power-law fracture length and permeability distributions characterized by exponents a and δ in zone I of figure 8. The channeling and homogenization lengths are neither defined.

7 Comparison with scale effects observed in the field

Permeability scalings observed in the field [59, 60] give a first insight into the relevant models of equivalent permeability and in turn in the relevant model of fracture length and permeability distributions. Although the compilation of Clauser does not respect the site uniqueness, it gives some insights into the application of this study. Data display a variation of the network permeability over three orders of magnitude, from the laboratory to the regional scale. There is a regional scale ($L>100$ m ... 1 km), at which the system permeability becomes constant. These qualitative observations suggest at least a selection of the potential relevant networks models presented here. The models that display the same type of permeability scale effects are the models of lognormal permeability distribution without correlation between fracture length and permeability for a in the interval [1,3], and the models of lognormal permeability distribution with correlation for a in the interval [2,3]. Figure 9 presents fits of Clauser's data by models of lognormal fracture transmissivity distribution with and without correlation. These data also rule out the power-law fracture transmissivity model.

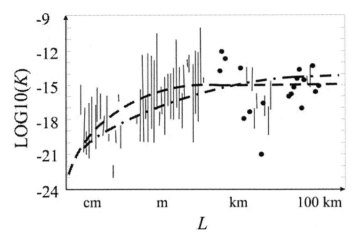

Fig. 9: Comparison of the field data of *Clauser* [1992] (points and vertical bars) with the permeability obtained on two models of lognormal fracture transmissivity distribution with and without correlation which are represented by the solid and dashed lines respectively. The model without correlation was fitted for a=2.7 and b=3.5 (dashed line), whereas the model with correlation was fitted for a=2.2 and b=0.8 (dash-dotted line).

We note finally that the permeability increase with scale is an effect of both length and permeability distributions. Other models based on only length or permeability distribution lead to a decreasing or a constant permeability. The models proposed here are more generally the only isotropic bi-dimensional models that produce a permeability increase with scale.

8 Conclusion

In this paper we discuss the consequences of the multi-scale nature of fault networks on flow property. The fact that fractures exists at all scales, from micro-cracks to pluri-kilometric faults is a challenging issue for defining average flow properties of fracture networks. We first claim that most of the fundamental effects are due to two basic properties of fracture networks: the fracture-length distribution which quantifies the scaling nature of fractures, and the transmissivity distribution per fracture. The fracture length distribution is reasonably modeled by a power law whose exponent a quantifies the scaling nature of fracture networks. It has been measured on 3D fracture outcrops with values ranging from 2 to 3. The transmissivity distribution is much less known considering the difficulty to measure transmissivity in hydrogeological sites. Lognormal functions or power laws are classically used as transmissivity distribution models, but this has to be taken as a conjecture.

The flow model is highly dependent on the scaling property of fracture networks, and more specifically on the power-law length exponent. The two end-member models are the percolation-like model, where all fractures have a length much smaller than the system size, and the single-fracture model where the fracture organization comes to consider only a few large fractures cross-cutting the system. In between, the probability of occurrence of large fractures (i.e. larger than the studied system) increases with scale such as systems are always connected above a critical connecting scale which depends on the power-law exponent a and on the fracture density.

Permeability as well as other flow properties is intimately related to the flow structure that we can characterize by three main length scales which depend on the fracture length distribution and on the transmissivity distribution per fracture (fig.10):

- the connection length, which define the crossover scale at which some networks shift from disconnected to connected,
- the channeling scale, which corresponds to a transition in the flow structure below which the flow is channeled into one main path, and above which it is distributed into several paths,
- and the homogenization scale, above which the network equivalent permeability becomes constant.

In some cases of fracture and permeability structure, these characteristic scales can be nil or infinite. may not exist depending on the. For instance, the channeling scale does not exist (or is infinite) for small cracks at percolation threshold, or for a finite number of infinite fractures. Likewise there are cases where the homogenization scale is infinite (fig.8 and 10), i.e. the permeability keeps going on increasing with scale.

Figure 10 summarizes these scaling effects on flow structure and on permeability.

a- Flow structure

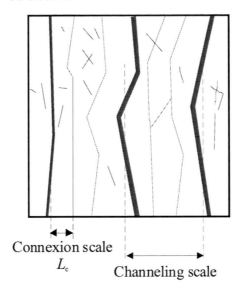

Connexion scale
L_c
Channeling scale

b- Permeability scale effects

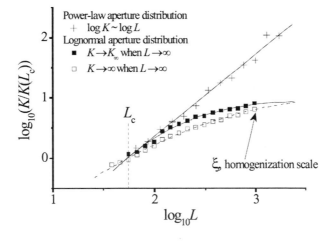

Figure 10: (a) Sketch of the flow structures with two characteristic length scales: the connecting scale L_c, and the channeling scale. (b) Scaling of the equivalent permeability calculated for three different transmissivity (or aperture) models. The homogenization scale ξ is defined as the scale above which the permeability becomes constant. It does not exist for the power-law transmissivity model represented by small black crosses.

Bibliography

[1] Weinrib, A., Long-ranged correlated percolation. *Physical Review B*, 1984. 29, 1, pp. 387-395.

[2] Prakash, S., et al., Structural and dynamical properties of long-ranged correlated percolation. *Physical Review A*, 1992. 46, 4, pp. R1724-1727.

[3] Schmittbuhl, J., J. Vilotte, and S. Roux, Percolation Through Self-Affine Surfaces. *Journal of Physics a-Mathematical and General*, 1993. 26, 22.

[4] Sahimi, M. and S. Mukhopdihyay, Scaling properties of a percolation model with long-ranged correlations. *Physical Review E*, 1996. 54, 4, pp. 3870-3880.

[5] Snow, D.T., Anisotropic permeability of fractured media. *Water Resources Research*, 1969, 5, pp. 1273-1289.

[6] Kiraly, L., Rapport sur l'état actuel des connaissances dans le domaine des caractères physiques des roches karstiques, in Hydrogeology of karstic terrains, A. Burger and L. Dubertret, Editors. 1975, International association of hydrogeologists: Paris.

[7] Stauffer, D. and A. Aharony, Introduction to percolation theory, second edition. 1992, Bristol: Taylor and Francis.

[8] Tchalenko, J.S., Similarities between shear zones of different magnitudes. *Geological Society of America Bulletin*, 1970. 81, pp. 1625-1640.

[9] Mandelbrot, B.B., The fractal geometry of Nature. 1982, New-York: W.H. Freeman. 468.

[10] Allègre, C.J., J.L. Lemouel, and A. Provost, Scaling rules in rock fracture and possible applications for earthquake prediction. *Nature*, 1982. 297, 47-49.

[11] King, G., The accommodation of large strains in the upper lithosphere of the earth and other solids by self-similar fault-system. *Pure and Applied Geophysics*, 1983. 12, pp. 761-915.

[12] Turcotte, D.L., A fractal model for crustal deformation. *Tectonophysics*, 1986. 132, pp. 261-269.

[13] Davy, P., et al., Localization and fault growth in brittle-ductile systems. Implications to deformations of the continental lithosphere. *Journal Geophysical Research*, 1995. 100, pp. 6281-6294.

[14] Bonnet, E., et al., Scaling of fracture systems in geological media. *Reviews of Geophysics*, 2001. 39, 3, pp. 347-384.

[15] Bour, O., et al., A statistical scaling model for fracture network geometry, with validation on a multi-scale mapping of a joint network (Hornelen Basin, Norway). *J. Geophys. Res.*, 2002. 107, 10.1029/2001JB000176.

[16] Davy, P., A. Sornette, and D. Sornette, Some consequences of a proposed fractal nature of continental faulting. *Nature*, 1990. 348, pp. 56-58.

[17] Heffer, K.J. and T.G. Bevan. Scaling relationships in natural fractures: data, theory, and application. in 2nd European Petroleum Conference. 1990.

[18] Yielding, G., J.J. Walsh, and J. Watterson, The prediction of small scale faulting in reservoirs. *First Break*, 1992. 10, pp. 449-460.

[19] Scholz, C.H., et al., Fault growth and fault scaling laws : Preliminary results. *Journal of Geophysical Research*, 1993. 98, pp. 21951-21961.

[20] Castaing, C., et al., Scaling relationships in intraplate fracture systems related to Red Sea Rifting. *Tectonophysics*, 1996. 261, pp. 291-314.

[21] Odling, N.E., Scaling and connectivity of joint systems in sandstones from western Norway. *Journal of Structural Geology*, 1997. 19, 10, pp. 1257-1271.

[22] Segall, P. and D.D. Pollard, Joint formation in granitic rock of the Sierra Nevada. *Geological Society of American Bulletin*, 1983. 94, pp. 563-575.

[23] Villemin, T. and C. Sunwoo, Distribution logarithmique des rejets et longueurs de failles: Exemple du bassin houiller lorrain. *Compte-rendus de l'Académie des Sciences*, Paris, 1987. 305, pp. 1309-1312.

[24] Childs, C., J.J. Walsh, and J. Watterson, A method for estimation of the density of fault displacements below the limits of seismic resolution in reservoir formations, A.T.B.e. al, Editor. 1990, Graham and Trotman: London. p. 309-318.

[25] Scholz, C.H. and P.A. Cowie, Determination of total strain from faulting using slip measurements. *Nature*, 1990. 346, pp. 837-839.

[26] Davy, P., On the frequency-length distribution of the San Andreas fault system. *Journal of Geophysical Research*, 1993. 98, 12, pp. 12,141-12,151.

[27] Cowie, P.A., D. Sornette, and C. Vanneste, Multifractal scaling properties of a growing fault population. *Geophysical Journal International*, 1995. 122, pp. 457-469.

[28] Berkowitz, B., et al., Scaling of fracture connectivity in geological formations. *Geophysical Research Letters*, 2000. 27, 14.

[29] Bour, O. and P. Davy, Clustering and size distributions of Fault Patterns : theory and measurements. *Geophys. Res. Lett.*, 1999. 26, pp. 2001-2004.

[30] Darcel, C., O. Bour, and P. Davy, Cross-correlation between length and position in real fracture networks. *Geophys. Res. Lett.*, 2003.

[31] Trimmer, D., et al., Effect of Pressure and Stress on Water Transport in Intact and Fractured Gabro and Granite. *Journal of Geophysical Research*, 1980. 85, B12, pp. 7059-7071.

[32] National Research Council, Rock Fractures and Fluid Flow. 1996, Washington, D.C.: National Academy Press.

[33] Robinson, P.C., Connectivity of fracture systems-A percolation theory approach. *J. Phys. A*, 1983. 16, 3, pp. 605-614.

[34] Hestir, K. and J.C.S. Long, Analytical Expression for the Permeability of Random Two-Dimensional Poisson Fracture Networks Based on Regular Lattice Percolation and Equivalent Media Theories. *Journal of Geophysical Research*, 1990. 95, B13, pp. 565-581.

[35] Berkowitz, B., Analysis of fracture network connectivity using percolation theory. *Mathematical Geology*, 1995. 27, 4, pp. 467-483.

[36] Watanabe, K. and H. Takahashi, Fractal geometry characterization of geothermal reservoir fracture networks. *Journal of Geophysical Research*, 1995. 100, B1, pp. 521-528.

[37] Bour, O. and P. Davy, Connectivity of random fault networks following a power law fault length distribution. *Water Resources Research*, 1997. 33, 7, pp. 1567-1583.

[38] Bour, O. and P. Davy, On the connectivity of three dimensional fault networks. *Water Resources Research*, 1998. 34, 10, pp. 2611-2622.

[39] de Dreuzy, J.R., P. Davy, and O. Bour, Percolation threshold of 3D random ellipses with widely-scattered distributions of eccentricity and size. *Physical Review E*, 2000. 62, 5, pp. 5948-5952.

[40] de Dreuzy, J.R., P. Davy, and O. Bour, Hydraulic properties of two-dimensional random fracture networks following a power law length distribution: 2-Permeability of networks based on log-normal distribution of apertures. *Water Resources Research*, 2001, in press.

[41] de Dreuzy, J.R., P. Davy, and O. Bour, Hydraulic properties of two-dimensional random fracture networks following a power law length distribution: 1-Effecive connectivity. *Water Resources Research*, 2001, in press.

[42] de Dreuzy, J.R., P. Davy, and O. Bour, Permeability of 2D fracture networks with power-law distributions of length and aperture. Submitted to *Water Resources Research*, 2001.

[43] Hsieh, P.A., Scale effects in fluid flow through fractured geological media, in Scale dependence and scale invariance in hydrology. 1998, Cambridge University Press. p. 335-353.

[44] Witherspoon, P.A., et al., Validity of Cubic Law for fluid flow in a deformable rock fracture. *Water Resources Research*, 1980. 16, 6, pp. 1016-1024.

[45] Pyrak-Nolte, L.J., N.G. Cook, and D.D. Nolte, Fluid percolation through single fractures. *Geophysical Research Letters*, 1988. 15, 11, pp. 1247-1250.

[46] Renshaw, C.E., On the relationship between mechanical and hydraulic apertures in rough-walled fractures. *Journal of Geophysical Research*, 1995. 100, B12, pp. 629-636.

[47] Ge, S., A governing equation for fluid flow in rough fractures. *Water Resources Research*, 1997. 33, 1, pp. 53-61.

[48] Oron, A.P. and B. Berkowitz, Flow in rock fractures: The local cubic law assumption reexamined. *Water Resources Research*, 1998. 34, 11, pp. 2811-2825.

[49] Dijk, P. and B. Berkowitz, Three-dimensional flow measurements in rock fractures. *Water Resources Research*, 1999. 35, 12.

[50] Barton, C.C. and P.A. Hsieh, Physical and Hydrologic flow properties of fractures, Field trip Guide. 1989, AGU, Washington D.C.

[51] Wong, T.F., J.T. Fredrich, and G.D. Gwanmesia, Crack aperture statistics and pore space fractal geometry of Westerly granite and Rutland quartzite : implications for an elastic contact model of rock compressibility. *Journal of Geophysical Research*, 1989. 94, pp. 10267-10278.

[52] Belfield, W.C., Multifractal characteristics of natural fracture apertures. *Journal of Structural Geology*, 1994. 21, 24, pp. 2641-2644.

[53] Dverstop, B. and J. Andersson, Application of the Discrete Fracture Network Concept With Filed Data: Possibilities of Model Calibration and Validation. *Water Resources Research*, 1989. 25, 3, pp. 540-550.

[54] Cacas, M.C., et al., Modeling fracture flow with a stochastic discrete fracture network: calibration and validation. 1. The flow model. *Water Resources Research*, 1990. 26, 3, pp. 479-489.

[55] Tsang, Y.W., et al., Tracer transport in a stochastic continuum model of fractured media. *Water Resources Research*, 1996. 32, 10, pp. 3077-3092.

[56] Neuman, S.P., Generalized Scaling of Permeabilities. *Geophysical Research Letters*, 1994. 21, 5, pp. 349-352.

[57] Charlaix, E., E. Guyon, and S. Roux, Permeability of a Random Array of Fractures of Widely Varying Apertures. *Transport in Porous Media*, 1987. 2, pp. 31-43.

[58] Friedman, S.P. and N.A. Seaton, Critical path analysis of the relationship between permeability and electrical conductivity of 3-dimensional pore networks. *Water Resour. Res.*, 1998. 34, pp. 1703-1710.

[59] Clauser, C., Permeability of crystalline rock. *Eos Trans. AGU*, 1992. 73, 21, pp. 237-238.

[60] Schulze-Makuch, D. and D.S. Cherkauer, Method Developed for Extrapoling Scale Behavior. *Eos Trans. AGU*, 1997. 78, 1, pp. 3.